Johann Caspar Lavater

Physiognomy

Johann Caspar Lavater

Physiognomy

ISBN/EAN: 9783337400026

Printed in Europe, USA, Canada, Australia, Japan

Cover: Foto ©berggeist007 / pixelio.de

More available books at **www.hansebooks.com**

PHYSIOGNOMY;

OR,

THE CORRESPONDING ANALOGY BETWEEN THE CONFORMATION OF THE FEATURES

AND THE

RULING PASSIONS OF THE MIND:

BEING

A COMPLETE EPITOME OF THE ORIGINAL WORK

OF

J. C. LAVATER.

New Edition.—Illustrated.

"Physiognomy is reading the handwriting of nature upon the human countenance."

LONDON: WILLIAM TEGG.
1866.

PREFACE.

THERE is undoubtedly no subject in the science of Natural History more curious, entertaining, and instructive to the human race in general, than that which respects the variety of complexion and figure among mankind. Though much has been written to point out the sources from whence these varieties arise, and to investigate the causes which certainly produce them, yet hitherto but little accurate information has been derived from the most arduous and laborious researches of the first abstract philosophers of the age.

The same thing has happened to Physiognomy as to Astronomy: they have both been degraded and disgraced by the intrigues and artifices of interested knavery. The first has been connected to palmistry by a notorious set of dusky impostors, who, roving up and down in the world, have made a prey of every credulous person they could meet with ; and the other has been travestied in the art of divining future events. Hence have arisen conjurers; the most notorious of which, combining the whole together, have not only found admirers in the less informed ages of the world, but are even daring enough yet, at the latter end of the eighteenth century, to hold up their arguments in defiance of experimental philosophy.

Confused and sophisticated with falsehoods, termed occult reasonings, the noble science of Physiognomy has been neglected for near a century, and deemed by the judicious a mere farcical contrivance to fleece the pockets and disturb the brains of the unwary. Thus even those who have suspected there might be some rational grounds to build hypothesis upon, have been fearful of venturing to appear even in the slender form of an essay.

From an accurate survey of all that has hitherto been written upon this subject by the soberest authors of the preceding age, it will appear that very little knowledge of man has been derived; and the falsehoods and errors with which their writings abound, are daily becoming more evident. Those systems which were established on authorities so extremely weak, are now falling into that contempt and neglect which must necessarily await every mode of reasoning whose axioms are not founded on obvious and derivative facts, and supported by physical causes.

The noble ardour for discovering and investigating the connection between the inward and outward operations of nature in man gave rise, in a neighbouring nation, to a splendid and expensive work,* an epitome of which is here offered to the public, arranged (the Editor hopes he may say without presumption) with more order and method, and divested of the numerous repetitions which the worthy and amiable, but too often rhapsodical LAVATER, in the warmth of a disinterested love of mankind, introduces at every turn.

In the present state of our knowledge, a systematical view of the physiognomical science can hardly be expected: a collection of observations arranged but with little attention to method, is all the industrious Lavater promises, and all we can reasonably expect. However, he furnishes us with an instance how much may be accomplished, even by an individual, in a subject replete with difficulties, when genius and judgment are aided by labour, and when the object is pursued with a steady regard to truth and veracity. However, it is not the Editor's intention to enter into any panegyric upon the labours of M. Lavater: the public will ever judge for themselves, and pay the tribute of applause where it is due.

* Published by William Tegg.

 PREFACE.

To preserve the spirit of Lavater's reasoning, inspire the enthusiasm of his feelings, and the sublimity of his conceptions, has been the endeavour of the Editor of the present volume, within the small compass of which, he flatters himself, he has concentrated, as in a focus, all the discoveries and truths contained in the original work.

1865.

CONTENTS.

CONTENTS.

PHYSIOGNOMY.

CHAPTER I.

INTRODUCTION.

Physiognomy a science—The truth of Physiognomy—The advantages of Physiognomy—Its disadvantages—The ease and difficulty of studying Physiognomy—A word concerning the Author.

It has been asserted by thousands, that "though there may be some truth in physiognomy, still it never can be a science." These assertions will be repeated, how clearly soever their objections may be answered, and however little they may have to reply. Physiognomy is as capable of becoming a science as any one of the sciences, mathematics excepted. It is a branch of the physical art, and includes theology and the belles lettres. Like these, it may to a certain extent be reduced to rule, and acquire an appropriate character by which it may be taught.

Whenever truth or knowledge is explained by fixed principles, it becomes scientific, so far as it can be imparted by words, lines, rules, and definitions. The question will stand simply thus : whether it be possible to explain the undeniable striking differences which exist between human faces and forms, not by obscure and confused conceptions, but by certain characters, signs, and expressions? Whether these signs can communicate the strength and weakness, health and sickness, of the body; the folly and wisdom, the magnanimity and meanness, the virtue and vice, of the mind? This is the only thing to be decided; and he who, instead of investigating the question, should continue to declaim against it, must either be deficient in the love of truth, or in logical reasoning.

The experimental philosopher can only proceed with his discoveries to a certain extent; only can communicate them by words; can only say, "Such and such are my experiments, such my remarks, such is the number of them, and such are the inferences I draw: pursue the track that I have explored." Yet will he not be unable, sometimes, to say thus much? Will not his active mind make a thousand remarks which he will want the power to communicate? Will not his eye penetrate recesses which he shall be unable to discover to that feebler vision that cannot discover for itself? Is any science brought to perfection at the moment of its birth? Does not genius continually, with eagle eye and flight, anticipate centuries? How long did the world wait for Wolf? Who, among the moderns, is more scientific than Bonnet? Who more accurately distinguishes falsehood from truth? Yet to whom would he be able to communicate his sudden perception of the truth; the result or resources of those numerous, small, indescribable, rapid, profound remarks? To whom could he impart these by signs, tones, images, and rules? Is it not the same with physic, theology, and all the arts and sciences? Is it not the same with painting, at once the mother and daughter of physiognomy?

How infinitely does he, who is painter or poet born, soar beyond all written rule! But must he who pos-

sesses feelings and power which are not to be reduced
to rule, be pronounced unscientific ?   So, physiognomical
truth may, to a certain degree, be defined, communicated
by signs and words, as a science.   This is the look of
contempt, this of innocence.   Where such signs are, such
and such properties reside.

There can be no doubt of the truth of physiognomy.
All countenances, all forms, all created beings, are not
only different from each other in their classes, races, and
kinds, but are also individually distinct.   Each being
differs from every other being of its species.   However
generally known, it is a truth the most important to our
purpose, and necessary to repeat, that "there is no rose
perfectly similar to another rose, no egg to an egg, no
eel to an eel, no lion to a lion, no eagle to an eagle, no
man to a man."

Confining this proposition to man only, it is the first,
the most profound, most secure and unshaken founda-
tion-stone of physiognomy, that, however intimate the
analogy and similarity of the innumerable forms of men,
no two men can be found who, brought together and
accurately compared, will not appear to be very remarkably different.   Nor is it less incontrovertible that it is
equally impossible to find two minds, as two countenances, which perfectly resemble each other.

Considerations like these will be sufficient to make
it received as a truth not requiring farther demonstration, that there must be a certain native analogy between
the external varieties of the countenance and form,
and the internal varieties of the mind.   Anger renders
the muscles protuberant; and shall not, therefore, an
angry mind and protuberant muscles be considered as
cause and effect?

After repeated observation, that an active and vivid eye, and an active and acute wit, are frequently found in the same person, shall it be supposed that there is no relation between the active eye and the active mind? Is this the effect of accident? Ought it not rather to be considered as sympathy, an interchangeable and instantaneous effect, when we perceive that, at the very moment the understanding is most acute and penetrating, and the wit the most lively, the motion and fire of the eye undergo, at that moment, the most visible alteration?

But all this is denied by those who oppose the truth of the science of physiognomy. Truth, according to them, is ever at variance with herself; eternal order is degraded to a juggler, whose purpose it is to deceive.

Calm reason revolts when it is asserted that the strong man may appear perfectly like the weak, the man in full health like another in the last stage of a consumption, or that the rash and irascible resemble the cold and phlegmatic. It revolts to hear it affirmed that joy and grief, pleasure and pain, love and hatred, all exhibit themselves under the same traits—that is to say, under no traits whatever—on the exterior of man. Yet such are the assertions of those who maintain that physiognomy is a chimerical science. They overturn all that order and combination by which Eternal Wisdom so highly astonishes and delights the understanding. It cannot be too emphatically repeated, that blind chance and arbitrary disorder constitute the philosophy of fools, and that they are the bane of natural knowledge, philosophy, and religion. Entirely to banish such a system is the duty of the true inquirer, the sage, and the divine.

It is indisputable that all men, absolutely all men, estimate all things whatever by their physiognomy, their exterior temporary superficies. By viewing these on every occasion, they draw their conclusions concerning their internal properties. What merchant, if he be unacquainted with the person of whom he purchases, does not estimate his wares by the physiognomy or appearance of those wares? If he purchase of a distant correspondent, what other means does he use in judging whether they are or are not equal to his expectation? Is not his judgment determined by the colour, the fineness, the superficies, the exterior, the physiognomy? Does he not judge money by its physiognomy? Why does he take one guinea and reject another? Why weigh a third in his hand? Does he not determine according to its colour, or impression, its outside, its physiognomy? If a stranger enter his shop as a buyer or seller, will he not observe him? Will he not draw conclusions from his countenance? Will he not, almost before he is out of hearing, pronounce some opinion of him, and say, "This man has an honest look—this man has a pleasing or forbidding countenance?" What is it to the purpose whether his judgment be right or wrong? He judges; and though not wholly, he depends, in part, upon the exterior form, and thence draws inferences concerning the mind.

The farmer, walking through his grounds, regulates his future expectations by the colour, the size, the growth, the exterior; that is to say, by the physiognomy of the bloom, the stalk, or the ear of his corn, the stem and shoots of his vine-tree. "This ear of corn is blighted —that wood is full of sap—this will grow, that not," affirms he at the first or second glance. "Though these vine-shoots look well, they will bear but few grapes." And wherefore? He remarks in their appearance, as the physiognomist in the countenances of shallow men, the want of native energy. Does he not judge by the exterior?

Does not the physician pay more attention to the physiognomy of the sick than to all the accounts that are brought him concerning his patient? Zimmerman, among the living, may be brought as a proof of the great perfection at which this kind of judgment is arrived; and, among the dead, Kempf, whose son has written a treatise on temperament.

. I will say nothing of the painter, as his art too evidently reproves the childish and arrogant prejudices of those who pretend to disbelieve physiognomy. The traveller, the philanthropist, the misanthropist, the lover, (and who not?) all act according to their feelings and decisions, true or false, confused or clear, concerning physiognomy. These feelings, these decisions, excite compassion, disgust, joy, love, hatred, suspicion, confidence, reserve, or benevolence.

By what rule do we judge of the sky but by its physiognomy? No food, not a glass of wine or beer, nor a cup of coffee or tea, comes to table, which is not judged by its physiognomy, its exterior, and of which we do not then deduce some conclusion respecting its interior good or bad properties. Is not all nature physiognomy, superficies and contents, body and spirit, exterior effect and internal power, invisible beginning and visible ending?

Physiognomy, whether understood in its most extensive or confined signification, is the origin of all human decisions, efforts, actions, expectations, fears, and hopes; of all pleasing and unpleasing sensations, which are occasioned by external objects. From the cradle to the grave, in all conditions and ages, throughout all nations, from Adam to the last existing man, from the worm we tread on to the most sublime of philosophers, physiognomy is the origin of all we do and suffer.

Every insect is acquainted with its friend and its foe; each child loves and fears, although it knows not why. Physiognomy is the cause: nor is there a man to be found on earth who is not daily influenced by physiognomy; not a man who cannot figure to himself a countenance which shall to him appear exceedingly lovely or exceedingly hateful; not a man who does not, more or less, the first time he is in company with a stranger, observe, estimate, compare, and judge of him according to appearances, although he might never have heard of the word or thing called physiognomy; not a man who does not judge of all things that pass through his hands by their physiognomy, that is, their internal worth by their external appearance.

The act of dissimulation itself, which is adduced as so insuperable an objection to the truth of physiognomy, is founded upon physiognomy. Why does the hypocrite assume the appearance of an honest man, but because that he is convinced, though not perhaps from any systematic reflection, that all eyes are acquainted with the characteristic mark of honesty?

What judge, wise or unwise, whether the criminal confess or deny the fact, does not sometimes in this sense decide from appearances? Who can, is, or ought to be absolutely indifferent to the exterior of persons brought before him to be judged? What king would choose a minister without examining his exterior, secretly at least, and to a certain extent? An officer will not enlist a soldier without thus examining his appearance, putting his height out of the question. What master or mistress of a family will choose a servant without considering the exterior? No matter that their judgment may or may not be just, or that it may be exercised unconsciously.

I am weary of citing such numerous instances, which are so continually before our eyes, to prove that men, tacitly and unanimously, confess the influence which physiognomy has over their sensations and actions. I feel disgust at being obliged to write thus, in order to convince the learned of truths which lie within the reach of every child.

Let him see who has eyes to see; but should the light, by being brought too close to his eyes, produce frenzy, he may burn himself by endeavouring to extinguish the torch of truth. I am not fond of using such expressions; but I dare to do my duty, and my duty is boldly to declare that I believe myself certain of what I now and hereafter shall affirm; and that I think myself capable of convincing all lovers of truth, by principles which are in themselves incontrovertible. It is also necessary to confute the pretensions of certain literary despots, and to compel them to be more cautious in their decisions. It is therefore proved, it being an eternal and manifest truth, that, whether they are or are not sensible of it, all men are daily influenced by physiognomy; nay, there is not a living being which does not, at least after its manner, draw some inferences from the external to the internal; which does not judge concerning that which is not, by that which is apparent to the senses.

This universal though tacit confession, that the exterior, the visible, the superficies of objects, indicate their nature, their properties, and that every outward sign is the symbol of some inherent quality, I hold to be equally certain and important to the science of physiognomy.

When each apple, each apricot, has a physiognomy peculiar to itself, shall man, the lord of the earth, have none? The most simple and inanimate object has its characteristic exterior, by which it is not only distinguished as a species, but individually; and shall the first, noblest, best harmonized, and most beautiful being, be denied all characteristic?

Whatever may be objected against the truth and certainty of the science of physiognomy by the most illiterate or the most learned; how much soever he, who openly professes faith in this science, may be subject to ridicule, to philosophic pity and contempt; it still cannot be contested, that there is no subject, thus considered, more important, more worthy of observation, more interesting than man, nor any occupation superior to that of disclosing the beauties and perfections of human nature.

I shall now proceed to inquire into the *advantages* of physiognomy. Whether a more certain, more accurate, more extensive, and thereby a more perfect knowledge of man, be or be not profitable; whether it be or be not advantageous to gain a knowledge of internal qualities from external form and feature, is a question most deserving of inquiry. This may be classed first as a general question, Whether knowledge, its extension and increase, be of consequence to man?

Certain it is, that if a man has the power, faculties, and will to obtain wisdom, that he should exercise those faculties for the attainment of wisdom. How paradoxical are those proofs, that science and knowledge are detrimental to man, and that a rude state of ignorance is to be preferred to all that wisdom can teach! I here dare assert, that physiognomy has at least as many claims of essential advantage as are granted by men in general to other sciences.

With how much justice may we not grant precedency to that science which teaches the knowledge of men! What object is so important to man as man himself? What knowledge can more influence his happiness than the knowledge of himself? This advantageous knowledge is the peculiar province of physiognomy.

Whoever would wish perfect conviction of the advantages of physiognomy, let him imagine, but for a moment, that all physiognomical knowledge and sensation were lost to the world. What confusion, what uncertainty and absurdity must take place, in millions of instances, among the actions of men! How perpetual must be the vexation of the eternal uncertainty in all which we should have to transact with each other; and how infinitely would probability, which depends upon a multitude of circumstances more or less distinctly perceived, be weakened by this privation! From how vast a number of actions, by which men are honoured and benefited, must they then desist!

Mutual intercourse is the thing of most consequence to mankind who are destined to live in society. The knowledge of man is the soul of this intercourse, that which imparts animation to it, pleasure, and profit. Let the physiognomist observe varieties, make minute distinctions, establish signs, and invent words, to express these his remarks; form general abstract propositions; extend and improve physiognomical knowledge, language, and sensation; and thus will the uses and advantages of physiognomy progressively increase.

Physiognomy is a source of the purest, the most exalted sensations; an additional eye, wherewith to view the manifold proofs of Divine wisdom and goodness in the creation, and, while thus viewing unspeakable harmony and truth, to excite more ecstatic love for their adorable Author. Where the dark, inattentive sight of the inexperienced perceives nothing, there the practical view of the physiognomist discovers inexhaustible fountains of delight, endearing, moral, and spiritual. With secret delight, the philanthropic physiognomist discerns those internal motives which would otherwise be first revealed in the world to come. He distinguishes what is permanent in the character from what is habitual, and what is habitual from what is accidental. He, therefore, who reads man in this language, reads him most accurately.

To enumerate all the advantages of physiognomy would require a large treatise. The most indisputable, though the most important of these, its advantages, are those the painter acquires, who, if he be not a physiognomist, is nothing. The greatest is that of forming, conducting, and improving the human heart.

I shall now say something with respect to the *disadvantages* of physiognomy.

Methinks I hear some worthy man exclaim : " O thou, who hast ever hitherto lived the friend of religion and virtue ! what is thy present purpose ? What mischief shall not be wrought by this thy physiognomy ? Wilt thou teach man the unblessed art of judging his brother by the ambiguous expressions of his countenance ? Are there not already sufficient of censoriousness, scandal, and inspection into the failings of others ? Wilt thou teach man to read the secrets of the heart, the latent feelings, and the various errors of thought ?

"Thou dwellest upon the advantages of the science; sayest thou shalt teach men to contemplate the beauty of virtue, the hatefulness of vice, and by these means make them virtuous; and that thou inspirest us with an abhorrence of vice by obliging us to feel its external deformity. And what shall be the consequence? Shall it not be, that for the appearance, and not the reality of goodness, man shall wish to be good? that, vain as he already is, acting from the desire of praise, and wishing only to appear what he ought determinately to be, he will yet become more vain, and will court the praise of men, not by words and deeds alone, but by assumed looks and counterfeited forms? Oughtest thou not rather to weaken this already too powerful motive for human actions, and to strengthen a better; to turn the eyes inward, to teach actual improvement and silent innocence, instead of inducing him to reason on the outward fair expressions of goodness, or the hateful ones of wickedness?"

This is a heavy accusation, and with great appearance of truth. Yet how easy is defence to me, and how pleasant, when my opponent accuses me from motives of philanthropy, and not of splenetic dispute! The charge is twofold, censoriousness and vanity. I will answer these charges separately; and now proceed to reply to the first objection.

I teach no black art; no nostrum, the secret of which I might have concealed, which is a thousand times injurious for once that it is profitable, the discovery of which is therefore so difficult. I do but teach a science, the most general, the most palpable, with which all men are acquainted; and state my feelings, observations, and their consequences.

It ought never to be forgotten, that the very purport of outward expression is to teach what passes in the mind, and that to deprive man of this source of knowledge were to reduce him to utter ignorance; that every man is born with a certain portion of physiognomical sensation, as certainly as that every man who is not deformed is born with two eyes; that all men, in their intercourse with each other, form physiognomical decisions according as their judgment is more or less clear; that it is well known, though physiognomy were never to be reduced to a science, most men, in proportion as they have mingled with the world, derive some profit from their knowledge of mankind, even at the first glance, and that the same effects were produced long before this question was in agitation. Whether, therefore, to teach men to decide with more perspicuity and certainty, instead of confusedly; to judge clearly with refined sensations, instead of rudely and erroneously with sensations more gross; and, instead of suffering them to wander in the dark, and venture abortive and injurious judgments, to learn them by physiognomical experiments, by the rules of prudence and caution, and the sublime voice of philanthrophy, to mistrust, to be diffident and slow to pronounce, where they imagine they discover evil: whether this, I say, can be injurious, I leave the world to determine.

I think I may venture to affirm, that very few persons will, in consequence of this work, begin to judge ill of others who had not before been guilty of the practice.

The second objection to physiognomy is, that "it renders men vain, and teaches them to assume a plausible appearance." The men thou wouldst reform are not children who are good, and know that they are so; but men who must, from experience, learn to distinguish between good and evil; men who, to become perfect, must necessarily be taught their own various, and consequently their own beneficent, qualities. Let, therefore, the desire of obtaining approbation from the good, act in concert with the impulse to goodness. Let this be the ladder, or, if you please, the crutch, to support tottering virtue. Suffer men to feel that God has ever branded vice with deformity, and adorned virtue with inimitable beauty. Allow man to rejoice when he perceives that his countenance improves in proportion as his heart is ennobled: Inform him only that to be good from vain motives is not actual good, but vanity; that the ornaments of vanity will ever be inferior and ignoble; and that the dignified mien of virtue never can be truly attained but by the actual possession of virtue, unsullied by the leaven of vanity.

Let me now say a word or two as to the *ease* and *difficulties* attending the study of physiognomy. To learn the lowest, the least difficult of sciences, at first appears an arduous undertaking, when taught by words or books, and not reduced to actual practice. What numerous dangers and difficulties might be started against all the daily enterprises of men, were it not undeniable that they are performed with facility. How might not the possibility of making a watch, and still more a watch worn in a ring, or of sailing over the vast ocean, and of numberless other arts and inventions, be disputed, did we not behold them constantly practised? How many arguments might be urged against the practice of physic? and, though some of them be unanswerable, how many are the reverse?

It is not just too hastily to decide on the possible ease

or difficulty of any subject which we have not yet examined. The simplest may abound with difficulties to him who has not made frequent experiments; and, by frequent experiments, the most difficult may become easy.

Whoever possesses the slightest capacity for, and has once acquired the habit of, observation and comparison, should he see himself daily and incessantly surrounded by hosts of difficulties, yet he will certainly be able to make a progress. There is no study, however difficult, which may not be attained by perseverance and resolution.

We have men constantly before us. In the very smallest towns there is a continual influx and reflux of persons of various and opposite characters : among these, many are known to us without consulting physiognomy; and that they are patient or choleric, credulous or suspicious, wise or foolish, of moderate or weak capacity, we are convinced past contradiction. Their countenances are as widely various as their characters, and these varieties of countenances may each be as accurately drawn as their varieties of character may be described.

There are men with whom we have daily intercourse, and whose interests and ours are connected; be their dissimulation what it may, passion will frequently for a moment snatch off the mask, and give us a glance, at least a side-view, of their true form.

Has Nature bestowed on man the eye and ear, and yet made her language so difficult, or so entirely unintelligible? And not the eye and ear alone, but feeling, nerves, internal sensations, and yet has rendered the language of the superficies so confused, so obscure? She who has adapted sound to the ear, and the ear to sound;

she who has created light for the eye, and the eye for light; she who has taught man so soon to speak, and to understand speech; shall she have imparted innumerable traits and marks of secret inclinations, powers, and passions, accompanied by perception, sensation, and an impulse to interpret them to his advantage; and, after bestowing such strong incitements, shall she have denied him the possibility of quenching this his thirst of knowledge? She who has given him penetration to discover sciences still more profound, though of much inferior utility; who has taught him to trace out the paths and measure the curves of comets; who has put a telescope into his hand, that he may view the satellites of the planets, and has endowed him with the capability of calculating their eclipses through revolving ages; shall so kind a mother have denied her children—her truth-seeking pupils, her noble philanthropic offspring, who are so willing to admire and rejoice in the majesty of the Most High, viewing man his masterpiece—the power of reading the ever-present, ever-open book of the human countenance; of reading man, the most beautiful of all her works, the compendium of all things, the mirror of the Deity?

Awake! view man in all his infinite forms! Look, for thou mayest eternally learn; shake off thy sloth, and behold! Meditate on its importance; take resolution to thyself, and the most difficult shall become easy.

Let me now mention the *difficulties* attending this study. There is a peculiar circumstance attending the starting of difficulties. There are some who possess the particular gift of discovering and inventing difficulties, without number or limits, on the most common and easy subjects. I shall be brief on the innumerable difficulties

of physiognomy; because, it not being my intention to
cite them all in this place, the most important will occasionally be noticed and answered in the course of the
work. I have an additional motive to be brief, which
is, that most of these difficulties are included in the indescribable minuteness of innumerable traits of character,
or the impossibility of seizing, expressing, and analyzing
certain sensations and observations.

Nothing can be more certain than that the smallest
shades, which are scarcely discernible to an inexperiened
eye, frequently denote total opposition of character. How
wonderfully may the expression of countenance and
character be altered by a small inflexion or diminishing,
lengthening or sharpening, even though but of a hair's
breadth !

How difficult, how impossible, must this variety of
the same countenance, even in the most accurate of the
arts of imitation, render precision ! How often does it
happen that the seat of character is so hidden, so envel-
oped, so masked, that it can only be caught in certain,
and perhaps uncommon, positions of the countenance;
which will again be changed, and the signs all disappear,
before they have made any durable impression ! or, supposing the impression made, these distinguishing traits
may be so difficult to seize, that it shall be impossible to
paint, much less to engrave, or describe them by language.

It is with physiognomy as with all other objects of
taste, literal or figurative, of sense or of spirit. How
many thousand accidents, great and small, physical and
moral; how many secret incidents, alterations, passions;
how often will dress, position, light and shade, and innumerable discordant circumstances, show the countenance
so disadvantageously, or, to speak more properly, betray

the physiognomist into a false judgment on the true qualities of the countenance and character! How easily may these occasion him to overlook the essential traits of character, and form his judgment on what is wholly accidental! How surprisingly may the small-pox, during life, disfigure the countenance! How may it destroy, confuse, or render the most decisive traits imperceptible!

We will therefore grant the opposer of physiognomy all he can ask, although we do not live without hope that many of the difficulties shall be resolved, which at first appeared to the reader and to the author inexplicable.*

It is highly incumbent upon me that I should not lead my readers to expect more from me than I am able to perform. Whoever publishes a considerable work on physiognomy, gives his readers apparently to understand that he is much better acquainted with the subject than any of his contemporaries. Should an error escape him, he exposes himself to the severest ridicule; he is contemned, at least by those who do not read him, for pretensions which probably they suppose him to make, but which in reality he does not make.

The God of truth, and all who know me, will bear testimony, that from my whole soul I despise deceit, as I do all silly claims to superior wisdom and infallibility, which so many writers, by a thousand artifices, endeavour to make their readers imagine they possess.

First, therefore, I declare, what I have uniformly declared on all occasions, although the persons who speak of me and my works endeavour to conceal it from themselves and others, that I understand but little of

---

* The following lines, to the end of the Introduction, contain Mr. Lavater's own remarks on himself.

, physiognomy; that I have been, and continue daily to be, mistaken in my judgment; but these errors are the most natural and most certain means of correcting, confirming, and extending my knowledge.

It will probably not be disagreeable to many of my readers to be informed, in part, of the progress of my mind in this study.

Before I reached the twenty-fifth year of my age, there was nothing I should have supposed more improbable than that I should make the smallest inquiries concerning, much less that I should write a book on, physiognomy. I was neither inclined to read nor make the slightest observations on the subject. The extreme sensibility of my nerves occasioned me, however, to feel certain emotions at beholding certain countenances. I sometimes instinctively formed a judgment according to these first impressions, and was laughed at, ashamed, and became cautious. Years passed away before I again dared, impelled by similar impressions, to venture similar opinions. In the meantime I occasionally sketched the countenance of a friend, whom by chance I had lately been observing. I had, from my earliest youth, a propensity to drawing, and especially to drawing of portraits, although I had but little genius or perseverance. By this practice my latent feelings began partly to unfold themselves. The various proportions, similitudes, and varieties of the human countenance became more apparent. It has happened that, on two successive days, I have drawn two faces, the features of which had a remarkable resemblance. This awakened my attention; and my astonishment increased when I received certain proofs that these persons were as similar in character as in feature.

I was afterwards induced by M. Zimmerman, physician to the court of Hanover, to write my thoughts on this subject. I met with many opponents; and this opposition obliged me to make deeper and more laborious researches, till at length the present work on physiognomy was produced.

Here I must repeat the full conviction I feel, that my whole life would be insufficient to form any approach towards a perfect and consistent whole. It is a field too vast for me singly to till. I shall find various opportunities of confessing my deficiency in various branches of science, without which it is impossible to study physiognomy with that firmness and certainty which are requisite. I shall conclude by declaring, with unreserved candour, and wholly committing myself to the reader who is the friend of truth,—

That I have heard, from the weakest men, remarks on the human countenance more acute than those I had made; remarks which made mine appear trifling.

That I believe, were various other people to sketch countenances and write their observations, those I have hitherto made would soon become of little importance.

That I daily meet an hundred faces concerning which I am unable to pronounce any certain opinion.

That no man has any thing to fear from my inspection, as it is my endeavour to find good in man; nor are there any men in whom good is not to be found.

That since I have begun thus to observe mankind, my philanthropy is not diminished, but, I will venture to say, increased.

And that now (January, 1783), after ten years' daily study, I am not more convinced of the certainty of my own existence than of the truth of the science of physiognomy, or than that this truth may be demonstrated; and that I hold him to be a weak and simple person who shall affirm that the effects of the impressions made upon him by all possible human countenances are equal.

CHAPTER II.

On the nature of Man, which is the foundation of the science of Physiognomy—Difference between Physiognomy and Pathognomy.

MAN is the most perfect of all earthly creatures, the most imbued with the principles of life. Each particle of matter is an immensity, each leaf a world, each insect an inexplicable compendium. Who, then, shall enumerate the gradations between insect and man? In him all the powers of nature are united. He is the essence of creation. The son of earth, he is the earth's lord; the summary and central point of all existence, of all powers, and of all life, on that earth which he inhabits.

There are no organized beings with which we are acquainted, man alone excepted, in which are so wonderfully united these different kinds of life, the animal, the intellectual, and the moral. Each of these lives is the compendium of various faculties most wonderfully compounded and harmonized.

To know, to desire, to act, or accurately to observe and meditate, to perceive and to wish, to possess the power of motion and resistance—these, combined, constitute man an animal, intellectual, and moral being.

Endowed with these faculties, and with this triple life, man is in himself the most worthy subject of observation, as he likewise is himself the most worthy observer

In him each species of life is conspicuous; yet never can his properties be wholly known except by the aid of his external form, his body, his superficies. How spiritual, how incorporeal soever his internal essence may be, still is he only visible and conceivable from the harmony of his constituent parts. From these he is inseparable. He exists and moves in the body he inhabits as in his element. This threefold life, which man cannot be denied to possess, necessarily first becomes the subject of disquisition and research as it presents itself in the form of body, and in such of his faculties as are apparent to sense.

By such external appearances as affect the senses, all things are characterised; they are the foundations of all human knowledge. Man must wander in the darkest ignorance, equally with respect to himself and the objects that surround him, did he not become acquainted with their properties and powers by the aid of their externals; and had not each object a character peculiar to its nature and essence, which acquaints us with what it is, and enables us to distinguish it from what it is not.

We survey all bodies that appear to sight under a certain form and superficies; we behold those outlines traced which are the result of their organization. I hope I shall be pardoned the repetition of common-place truths, since on these is built the science of physiognomy, or the proper study of man.

The organization of man peculiarly distinguishes him from all other earthly beings; and his physiognomy, that is to say, his superficies and outlines of this organization, show him to be infinitely superior to all those visible beings by which he is surrounded. We are unacquainted with any form equally noble, equally majestic, with that of man; and in which so many kinds of life, so many powers, so many virtues of action and motion, unite as in a central point. With firm step he advances over the earth's surface, and with erect body he raises his head to heaven. He looks forward to infinitude; he acts with facility and swiftness inconceivable, and his motions are the most immediate and the most varied. By whom may their varieties be enumerated? He can at once both suffer and perform infinitely more than any other creature. He unites flexibility and fortitude, strength and dexterity, activity and rest. Of all creatures he can the soonest yield, and the longest resist. None resemble him in the variety and harmony of his powers. His faculties, like his form, are peculiar to himself.

The make and proportion of man, his superior height, capable of so many changes and such variety of motion, prove to the unprejudiced observer his superior eminent strength, and astonishing facility of action. The high excellence and physiological unity of human nature, are visible at the first glance. The head, especially the face and the formation of the firm parts, compared to the firm parts of other animals, convince the accurate observer who is capable of investigating truth, of the greatness and superiority of his intellectual qualities. The eye, the look, the cheeks, the mouth, the forehead, whether considered in a state of entire rest, or during their innumerable varieties of motion—in fine, whatever is understood by physiognomy—are the most expressive, the most convincing picture of interior sensation, desires, passions, will, and of all those properties which so much exalt moral above animal life.

Although the physiological, intellectual, and moral

life of man, with all their subordinate powers and their constituent parts, so eminently unite in one being; although these three kinds of life do not, like three distinct families, reside in separate parts or stories of the body, but coexist in one point, and by their combination form one whole; yet it is plain that each of these powers of life has its peculiar station where it more especially unfolds itself and acts.

It is beyond contradiction evident, that, though physiological or animal life displays itself through all the body, and especially through all the animal parts, yet it acts more conspicuously in the arm, from the shoulder to the ends of the fingers.

It is not less evident that intellectual life, or the power of the understanding and the mind, make themselves most apparent in the circumference and form of the solid parts of the head, especially the forehead; though they will discover themselves to an attentive and accurate eye in every part and point of the human body, by the congeniality and harmony of the various parts. Is there any occasion to prove that the power of thinking resides neither in the foot, in the hand, nor in the back, but in the head and in its internal parts?

The moral life of man particularly reveals itself in the lines, marks, and transitions of the countenance. His moral powers and desires; his irritability, sympathy, and antipathy; his facility of attracting or repelling the objects that surround him: these are all summed up in, and painted upon his countenance when at rest. When any passion is called into action, such passion is depicted by the motion of the muscles, and these motions are accompanied by a strong palpitation of the heart. If the countenance be tranquil, it always denotes tranquillity in the region of the heart and breast.

This threefold life of man, so intimately interwoven through his frame, is still capable of being studied in its different appropriate parts; and, did we live in a less depraved world, we should find sufficient data for the science of physiognomy.

The animal life, the lowest and most earthly, would discover itself from the rim of the belly to the organs of generation, which would become its central or focal point. The middle or moral life would be seated in the breast, and the heart would be its central point. The intellectual life, which of the three is supreme, would reside in the head, and have the eye for its centre. If we take the countenance as the representative and epitome of the three divisions, then will the forehead to the eyebrows be the mirror or image of the understanding; the nose and cheeks, the image of the moral and sensitive life; and the mouth and chin, the image of the animal life; while the eye will be to the whole as its summary and centre.

All that has been hitherto advanced is so clear, so well known, so universal, that we should blush to insist upon such common-place truths, were they not, first, the foundation on which we must build all we have to propose; and, again, had not these truths (can it be believed by futurity?) in this our age been so many thousand times mistaken and contested with the most inconceivable affectation.

The science of physiognomy, whether understood in the most enlarged or most confined sense, indubitably depends on these general and incontrovertible principles; yet, incontrovertible as they are, they have not been without their opponents. Men pretend to doubt of the most striking, the most convincing, the most self-evident truths; although, were these destroyed, neither truth nor knowledge would remain. They do not profess to doubt concerning the physiognomy of other natural objects; yet do they doubt the physiognomy of human nature—the first object the most worthy of contemplation, and the most animated the realms of nature contain.

We have already hinted to our readers, that they are to expect only fragments on physiognomy from us, and not a perfect system. However, what has been said may serve as a sketch for such a system. We shall conclude this chapter with showing the difference between *Physiognomy* and *Pathognomy*.

Physiognomy is the science or knowledge of the correspondence between the external and internal man, the visible superficies and the invisible contents. Physiognomy, opposed to pathognomy, is the knowledge of the signs of the powers and inclinations of men—pathognomy is the knowledge of the signs of the passions. Physiognomy therefore teaches the knowledge of character at rest, and pathognomy of character in motion. Character at rest, is taught by the form of the solid and the appearance of the moveable parts while at rest. Character impassioned, is manifested by the moveable parts in motion.

Physiognomy may be compared to the sum-total of the mind; pathognomy, to the interest which is the product of this sum-total. The former shows what man is in general, the latter what he becomes at particular moments; or, the one what he might be, the other what he is. The first is the root and stem of the second, the soil in which it is planted. Whoever believes the latter

and not the former, believes in fruit without a tree, in
corn without land.

CHAPTER III.

Signs of Bodily Strength and Weakness—Of Health and Sickness.

WE call that human body strong which can easily
alter other bodies without being easily altered itself.
The more immediately it can act, and the less immediately it can be acted upon, the greater is its strength; and
the weaker, the less it can act or withstand the action of
others. There is a tranquil strength, the essence of
which is immobility; and there is an active strength,
the essence of which is motion. The one has motion,
the other stability, in an extraordinary degree. There
is the strength of the rock and the elasticity of the
spring.

There is the Herculean strength of bones and' sinews;
thick, firm, compact, and immoveable as a pillar.

There are heroes less Herculean, less firm, sinewy,
large; less set, less rocky; who yet, when roused, when
opposed in their activity, will meet oppression with so
much strength, will resist weight with such elastic force,
as scarcely to be equalled by the most muscular strength.

The elephant has native, bony strength. Irritated or
not, he bears prodigious burdens, and crushes all on
which he treads. An irritated wasp has strength of a
totally different kind; but both have compactness for
their foundation, and especially the firmness of construction. All porosity destroys strength.

The strength, like the understanding of a man, is

discovered by its being more or less compact. The elasticity of a body has signs so remarkable, that they will not permit us to confound such body with one that is not elastic. How manifest are the varieties of strength between the foot of an elephant and a stag, a wasp and a fly!

Tranquil, firm strength, is shown in the proportions of the form, which ought rather to be short than long. In the thick neck, the broad shoulders, and the countenance, which, in a state of health, is rather bony than fleshy. In the short, compact, and knotty forehead; and especially when the *sinus frontales* are visible, but not too far projecting; flat in the middle, or suddenly indented, but not in smooth cavities. In horizontal eyebrows, situated near the eye. Deep eyes and steadfast look. In the broad, firm nose, bony near the forehead, especially in its straight angular outlines. In short, thick, curly hair of the head and beard; broad teeth, standing close to each other. In compact lips, of which the under rather projects than retreats. In the strong, prominent, broad chin. In the strong, projecting *os occipitis*. In the bass voice, the firm step, and in sitting still.

Elastic strength, the living power of irritability, must be discovered in the moment of action; and the firm signs must afterwards be abstracted when the irritated power is once more at rest. "This body, therefore, which at rest was capable of so little, acted and resisted so weakly, can, thus irritated, and with this degree of tension, become thus powerful." We shall find on inquiry that this strength, awakened by irritation, generally resides in thin, tall, but not very tall, and bony rather than muscular bodies; in bodies of dark or pale complexions; of rapid motion, joined with a certain kind of stiffness; of hasty and firm walk; of fixed penetrating look; and with open lips, but easily and accurately to be closed.

Signs of weakness are, disproportionate length of body; much flesh; little bone; extension; a tottering frame; a loose skin; round, obtuse, and particularly hollow outlines of the forehead and nose; smallness of nose and chin; little nostrils; the retreating chin; long, cylindrical neck; the walk very hasty or languid, without firmness of step; the timid aspect; closing eyelids; open mouth; long teeth; the jawbone long, but bent towards the ear; whiteness of complexion; teeth inclined to be yellow or green; fair, long, and tender hair; shrill voice.

I shall now proceed to consider Medicinal Semeiotics, or the signs of *health* and *sickness*. Not I, but an experienced physician, ought to write on the physiognomical and pathognomical semeiotica of health and sickness, and describe the physiological character of the body, and its propensities to this or that disorder. I am beyond description ignorant with respect to the nature of disorders and their signs: still may I, in consequence of the few observations I have made, declare with some certainty, by repeatedly examining the firm parts and outlines of the bodies and countenances of the sick, that it is not difficult to predict what are the diseases to which the man in health is most liable.

Of what infinite importance would such physiognomical semeiotics, or prognostics of possible or probable disorders, be, founded on the nature and form of the body! How essential were it, could the physician say to the healthy, "You naturally have, some time in your life,

to expect this or that disorder. Take the necessary pre-
cautions against such or such a disease. The virus of
the small-pox slumbers in your body, and may thus or
thus be put in motion : thus the hectic, thus the inter-
mittent, and thus the putrid fever." Oh, how worthy,
Zimmerman, would a treatise on physiognomical *diætetice*
(or regimen) be of thee !

Whoever shall read this author's work on *experience,*
will see how characteristically he describes various diseases which originate in the passions. Some quotations
from this work, which will justify my wish, and contain
the most valuable semeiotical remarks, cannot be unacceptable to the reader :—

"The observing mind examines the physiognomy of
the sick, the signs of which extend over the whole body;
but the progress and change of the disease is principally
to be found in the countenance and its parts. Sometimes the patient carries the marks of his disease; in
burning, bilious, and hectic fevers; in the chlorosis; the
common and black jaundice; in worm cases." I, who
know so little of physic, have several times discovered
the disease of the tape-worm in the countenance.

"In the *furor uterinus* the least observant can read
the disease. The more the countenance is changed in
burning fevers, the greater is the danger. A man whose
natural aspect is mild and calm, but who stares at me,
with a florid complexion, and wildness in his eyes, prog-
nosticates an approaching delirium. I have likewise
seen a look indescribably wild, accompanied by paleness,
when nature, in an inflammation of the lungs, was coming to a crisis, and the patient was becoming excessively
cold and frantic. The countenance relaxed, the lips
pale and hanging, in burning fevers, are bad symptoms,

as they denote great debility; and if the change and
decay of the countenance be sudden, the danger is great.
When the nose is pointed, the face of a lead colour, and
the lips livid, inflammation has produced gangrene.

"There is frequently something dangerous to be
observed in the countenance, which cannot be known
from other symptoms, and which yet is very significant.
Much is to be observed in the eyes. Boerhaave
examined the eyes of the patient with a magnifying glass,
that he might see if the blood entered the smaller
vessels. Hippocrates held that the avoiding of light,
involuntary tears, squinting, one eye less than the other,
the white of the eye inflamed, the small veins inclined
to be black, too much swelled, or too much sunken,
were each and all bad symptoms.

"The motion of the patient, and his position in bed,
ought likewise to be enumerated among the particular
symptoms of disease. The hand carried to the forehead,
waved, or groping in the air, scratching on the wall, and
pulling up the bed-clothes, are of this kind. The
position in bed is a very significant sign of the internal
situation of the patient, and therefore deserves every
attention. The more unusual the position is in any
inflammatory disease, the more certainly we may conclude that the anguish is great, and consequently the
danger. Hippocrates has described the position of the
sick in such cases with an accuracy that leaves nothing
to be desired. The best position in sickness is the usual
position in health."

I shall add some other remarks from this physician
and physiognomist, whose abilities are superior to envy,
ignorance, and quackery. "Swift was lean while he was
the prey of ambition, chagrin, and ill-temper; but, after

the loss of his understanding, he became fat." His description of envy, and its effects on the body, is incomparable :—" The effects of envy are visible, even in children. They become thin, and easily fall into consumptions. Envy takes away the appetite and sleep, and causes feverish motion ; it produces gloom, shortness of breath, impatience, restlessness, and a narrow chest. The good name of others, on which it seeks to avenge itself by slander, and feigned but not real contempt, hangs like the sword suspended by a hair over the head of envy, that continually wishes to torture others, and is itself continually on the rack. The laughing simpleton becomes disturbed as soon as envy, that worst of fiends, takes possession of him, and he perceives that he vainly labours to debase that merit which he cannot rival. His eyes roll, he knits his forehead, he becomes morose, peevish, and hangs his lips. There is, it is true, a kind of envy that arrives at old age. Envy in her dark cave, possessed by toothless furies, there hoards her poison, which, with infernal wickedness, she endeavours to eject over each worthy person and honourable act. She defends the cause of vice, endeavours to confound right and wrong, and vitally wounds the purest innocence."

CHAPTER IV.

Of the Congeniality of the Human Form.

THE same vital powers that make the heart beat, give motion to the finger ; that which roofs the skull, arches the finger-nail. Art is at variance with herself: not so Nature. Her creation is progressive. From the head to the back, from the shoulder to the arm, from the arm to the hand, and from the hand to the finger; from the root to the stem, the stem to the branch, the branch to the twig, the twig to the blossom and fruit, each depends on the other, and all on the root: each is similar in nature and form. There is a determinate effect of a determinate power. Through all nature each determinate power is productive only of such and such determinate effects. The finger of one body is not adapted to the hand of another body. Each part of an organized body is an image of the whole. The blood in the extremity of the finger has the character of the blood in the heart. The same congeniality is found in the nerves, in the bones. One spirit lives in all. Each member of the body is in proportion to that whole of which it is a part. As from the length of the smallest member, the smallest joint of the finger, the proportion of the whole, the length and breadth of the body, may be found; so also may the form of the whole from the form of each single part. When the head is long all is long, or round when the head is round, or square when it is square. One form, one mind, one root, appertain to all: therefore is each organized body so much a whole, that, without discord, destruction, or deformity, nothing can be added or diminished.

Every thing in man is progressive; every thing congenial; form, stature, complexion, hair, skin, veins, nerves, bones, voice, walk, manner, style, passion, love, hatred. One and the same spirit is manifest in all. He has 'a determinate sphere in which his powers and sensations are allowed, within which they may be freely exercised, but beyond which he cannot pass. Each countenance is indeed subject to momentary change, though not perceptible, even in its solid parts; but

D

these changes are all proportionate : each is measured, each proper and peculiar to the countenance in which it takes place. The capability of change is limited. Even that which is affected, assumed, imitated, heterogeneous, still has the properties of the individual originating in the nature of the whole, and is so definite that it is only possible in this, but in no other being.

I almost blush to repeat this in the present age. What, Posterity, wilt thou suppose, thus to see me so often obliged to demonstrate to pretended sages that nature makes no emendation? She labours from one to all. Hers is not disjointed organization nor mosaic work. The more there is of the mosaic in the works of artists, orators, or poets, the less are they natural; the less do they resemble the copious streams of the fountain ; the stem extending itself to the remotest branch.

The more there is of progression, the more there is of truth, power, and nature; the more extensive, general, durable, and noble is the effect. The designs of nature are the designs of a moment ; one form, one spirit, appear through the whole. Thus nature forms her least plant, and thus her most exalted man. I shall have effected nothing by my physiognomical labours, if I am not able to destroy that opinion, so tasteless, so unworthy of the age, so opposite to all sound philosophy, that nature patches up the features of various countenances, in order to make one perfect countenance; and I shall think them well rewarded, if the congeniality, uniformity, and agreement of human organization be so demonstrated, that he who shall deny it will be declared to deny the light of the sun at noonday.

The human body is a plant, each part of which has the character of the stem. Suffer me to repeat this con-

tinually, since this most evident of all things is continually controverted, among all ranks of men, in words,
deeds, books, and works of art. I, therefore, find the
greatest incongruities in the heads of the greatest masters.
I know no painter of whom I can say he has thoroughly
studied the harmony of the human outline, not even
Poussin—no, not even Raphael himself. Let any one
class the forms of their countenances, and compare them
with the forms of nature. Let him, for instance, draw
the outlines of their foreheads, and endeavour to find
similar outlines in nature, and he will find incongruities
which could not have been expected in such great
masters.

Chodowiecki, excepting the too great length and
extent, particularly of his human figures, perhaps had
the most exact feeling of congeniality in caricature; that
is to say, of the relative propriety of the deformed, the
humorous, or other characteristical members and features.
For as there is conformity and congeniality in the beautiful, so is there also in the deformed. Every cripple
has the distortion peculiar to himself, the effects of which
are extended to his whole body. In like manner, the
evil actions of the evil, and the good actions of the good,
have a conformity of character; at least they are all
tinged with this conformity of character.

Little as this seems to be remarked by poets and
painters, still is it the foundation of their art; for wherever emendation is visible, there admiration is at an end.
Why has no painter yet been pleased to place the blue
eye beside the brown one? Yet, absurd as this would
be, no less absurd are the incongruities continually en-
countered by the physiognomical eye—the nose of Venus
on the head of Madona. I have been assured by a man

of fashion, that at a masquerade, with only the aid of an artificial nose, he entirely concealed himself from the knowledge of all his acquaintance. So much does nature reject what does not appertain to herself.

I have never yet met with one Roman nose among a hundred circular foreheads in profile. In a hundred other square foreheads, I have scarcely found one in which there were not cavities and prominences. I never yet saw a perpendicular forehead with strongly arched features in the lower part of the countenance, the double chin excepted.

I meet no strong-bowed eyebrows combined with bony perpendicular countenances.

Wherever the forehead is projecting, so in general are the under lips, children excepted.

I have never seen gently arched, yet much retreating foreheads, combined with a short snub nose, which in profile is sharp and sunken.

A visible nearness of the nose to the eye, is always attended by a visible wideness between the nose and mouth.

A long covering of the teeth, or, in other words, a long space between the nose and mouth, always indicates small upper lips. Length of form and face is generally attended by well-drawn fleshy lips.

I shall at present produce but one more example, which will convince all who possess acute physiognomical sensation, how great is the harmony of all nature's forms, and how much she hates the incongruous.

Take two, three, or four shades of men remarkable for understanding; join the features so artificially that no defect shall appear as far as relates to the act of joining; that is, take the forehead of one, add the nose of a second, the mouth of a third, the chin of a fourth, and the result of this combination of the signs of wisdom shall be folly. Folly is, perhaps, nothing more than the emendation of some heterogeneous addition. "But let these four wise countenances be supposed congruous." Let them so be supposed, or as nearly so as possible, still their combination will produce the signs of folly.

Those, therefore, who maintain that conclusion cannot be drawn from a part, from a single section of the profile, to the whole, would be perfectly right if unarbitrary nature patched up countenances like arbitrary art; but so, she does not. Indeed, when a man, being born with understanding, becomes a fool, there expression of heterogeneousness is the consequence. Either the lower part of the countenance extends itself, or the eyes acquire a direction not conformable to the forehead, the mouth cannot remain closed, or the features of the countenance, in some other manner, lose their consistency: all becomes discord; and folly, in such a countenance, is very manifest. Let him who would study physiognomy study the relation of the constituent parts of the countenance: not having studied these, he has studied nothing.

He only is an accurate physiognomist, and has the true spirit of physiognomy, who possesses sense, feeling, and sympathetic proportion of the congeniality and harmony of nature; and who hath a similar sense and feeling for all emendations and additions of art and constraint. He is no physiognomist who doubts of the propriety, simplicity, and harmony of nature, or who has not this physiognomical essential; who supposes nature selects members to form a whole, as a compositor in a printing-office does letters to make up a word; who

can suppose the works of nature are the patchwork of a
harlequin jacket. Not the most insignificant of insects
is so compounded, much less man, the most perfect of
organized beings. He respires not the breath of wisdom
who doubts of this progression, continuity, and simplicity
of the structure of nature. He wants a general feeling
for the works of nature; consequently of art, the imitator of nature. I shall be pardoned this warmth. It is
necessary. The consequences are infinite, and extend to
all things. He has the master-key of truth who has
this sensation of the congeniality of nature, and, by
necessary induction, of the human form.

All imperfection in works of art, productions of the
mind, moral actions, errors in judgment; all scepticism,
infidelity, and ridicule of religion, naturally originate in
the want of this knowledge and sensation. He soars
above all doubt of the Divinity and Christ who hath
them, and who is conscious of this congeniality. He also
who, at first sight, thoroughly understands and feels the
congeniality of the human form, and that from the want
of this congeniality arises the difference observed between
the works of nature and of art, is superior to all doubt
concerning the truth and divinity of the human countenance.

Those who have this sense, this feeling, call it which
you please, will attribute that only, and nothing more,
to each countenance which it is capable of receiving.
They will consider each according to its kind, and will
as little seek to add a heterogeneous character as a
heterogeneous nose to the face. Such will only unfold
what nature is desirous of unfolding, give what nature
is capable of receiving, and take away that with which
nature would not be encumbered. They will perceive in

Plate 1
Albert Durer. 1.
Moncrif. 2.
Dr Johnson. 3.
Shakespeare. 4.
Sterne. 5.
S. Clarke. 6.

the child, pupil, friend, or wife, when any discordant trait of character makes its appearance; and will endeavour to restore the original congeniality, the equilibrium of character and impulse, by acting upon the still remaining harmony, by co-operating with the yet unimpaired essential powers. They will consider each sin, each vice, as destructive of this harmony; will feel how much each departure from truth in the human form, at least to eyes more penetrating than human eyes are, must be manifest, must distort, and must become displeasing to the Creator, by rendering it unlike his image. Who, therefore, can judge better of the works and actions of man; who less offend or be offended; who more clearly develop cause and effect, than the physiognomist, possessed of a full portion of this knowledge and sensation?

CHAPTER V.

Description of Plates I. *and* II.

WE shall occasionally introduce some figures, in order to support and elucidate those opinions and propositions which may be advanced. These plates refer to objects that have been already alluded to in the preceding pages.

Description of Plate I. *Number* I.

This is a boldly sketched portrait of ALBERT DURER. Whoever examines this countenance cannot but perceive in it the traits of fortitude, deep penetration, determined. perseverance, and inventive genius. At least, every one will acknowledge the truth of these observations when made.

Number 2.   FAP. MONCRIF.

There are few men capable of observation who will class this visage with the stupid. In the aspect, the eye, the nose especially, and the mouth, are proofs, not to be mistaken, of the accomplished gentleman and the man of taste.

Number 3.   DR. SAMUEL JOHNSON.

The most unpractised eye will easily discover in this sketch of Johnson, the acute, the comprehensive, the capacious mind, not easily deceived, and rather inclined to suspicion than credulity.

Number 4.   W. SHAKSPEARE.

How deficient must all outlines be! Among ten thousand can one be found that is exact? Where is the outline that can portray genius? Yet, who does not read in this outline, imperfect as it is, from pure physiognomical sensation, the clear, the capacious, the rapid mind, all conceiving, all embracing that, with equal swiftness and facility, imagines, creates, produces?

Number 5.   L. STERNE.

The most unpractised reader in physiognomy will not deny to this countenance all the keen, the searching penetration of wit, the most original fancy, full of fire, and the powers of invention. Who is so dull as not to view in this countenance somewhat of the spirit of poor Yorick?

Number 6.   S. CLARKE.

Perspicuity, benevolence, dignity, serenity, dispassionate meditation, the powers of conception and perseverance, are the most apparent characteristics of this

N° 1.
N° 2.
N° 3.
N° 4.
N° 5.
N° 6.

countenance. He who can hate such a face, must laboriously counteract all those physiognomical sensations with which he was born.

Description of Plate II.

Hitherto we have beheld nature in the most perfect of her productions : we must now view the reverse : we must proceed to contemplate her in her deformity. In this also, how intelligibly does she speak to the eyes of all at the first glance!

Number 1.

Who does not here read reason debased, and stupidity almost sunken to brutality? This eye, these wrinkles of a lowering forehead, this projecting mouth, the whole position of the head, do they not all denote manifest dulness and debility?

Number 2. A Fool.

From the small eyes in this figure, the open mouth, particularly from the under part of the countenance, no man whatever will expect penetration, reasoning, or wisdom.

Number 3.

True or false, nature or caricature, this countenance will, to the common sensations of all men, depict an inhuman and brutal character. It is impossible that brutality should be overlooked in the nose and mouth, or in the eye, though still it deserves to be called a human eye.

Number 4.

Let us proceed to the characters of passion, which are intelligible to every child; so that concerning these there can be no dispute, if we are in any degree acquainted with their language. The more violent the passion is, the more apparent are its signs. The effect of the stiller passions is to contract, and of the violent to distend the muscles. Every one will perceive in this countenance fear mingled with abhorrence.

Number 5.

No man will expect cheerfulness, tranquillity, content, strength of mind, and magnanimity, from this countenance. Fear and terror are here strongly marked.

Number 6.

Terror, heightened by native indocility of character, is here strongly marked.

Such examples might be produced without end; but to adduce some of the most decisive of the various classes is sufficient. We shall give some farther specimens hereafter.

CHAPTER VI.

The universal Excellence of the Form of Man.

EACH creature is indispensable in the immensity of God's creation; but each creature does not know it is thus indispensable. Of all earth's creatures, man alone rejoices in his indispensability. No man can render any other man dispensable. The place of no man can be supplied by another.

This belief of the indispensability and individuality of all men, and in our own metaphysical indispensability and individuality, is one of the unacknowledged, the noble fruits of physiognomy; a fruit pregnant with most precious seed, whence shall spring lenity and love. Oh, may posterity behold them flourish! may future ages repose under their shade! The most deformed, the most corrupt of men, is still indispensable in this world of God, and is more or less capable of knowing his own individuality and unsuppliable indispensability. The wickedest, the most deformed of men, is still more noble than the most beauteous and perfect animal. Contemplate, O man! what thy nature is, not what it might be, not what is wanting. Humanity, amid all its distortions, will ever remain wondrous humanity!

Incessantly might I repeat doctrines like this. Art thou better, more beauteous, nobler, than many others of thy fellow-creatures? If so, rejoice, and ascribe it not to thyself, but to Him who, from the same clay, formed one vessel for honour, another for dishonour; to Him who, without thy advice, without thy prayer, without any desert of thine, caused thee to be what thou art.

Yea, tò Him! "for what hast thou, O man! that thou didst not receive? Now, if thou didst receive, why dost thou glory as if thou hadst not received?" "Can the eye say to the hand, I have no need of thee?" "He that oppresseth the poor reproacheth his Maker." "God hath made of one blood all nations of men." Who more deeply, more internally, feels all these divine truths than the physiognomist? the true physiognomist, who is not merely a man of literature, a reader, a reviewer, an author, but—a man!

I am ready to acknowledge that the most humane physiognomist, he who so eagerly searches whatever is good, beautiful, and noble in nature; who delights in the *ideal;* who duly exercises, nourishes, refines his taste, with humanity more improved, more perfect, more holy; even he is in frequent danger, at least is frequently tempted, to turn from the common herd of depraved men—from the deformed, the foolish, the apes, the hypocrites, the vulgar of mankind; in danger of forgetting that these misshapen forms, these apes, these hypocrites, also are men; and that, notwithstanding all his imagined or his real excellence, all his noble feelings, the purity of his views, (and who has cause to boast of these?) all the firmness, the soundness of his reason, the feelings of his heart, the powers with which he is endowed, still he is very probably, from his own moral defects, in the eyes of his superior beings, in the eyes of his much more righteous brother, as distorted as the most ridiculous, most depraved moral or physical monster appears to be in his eyes.

Liable as we are to forget this, reminding is necessary both to the writer and reader of this work. Forget not that even the wisest of men are men. Forget not how much positve good may be found even in the worst, and that they are as necessary, as good in their place, as thou art. Are they not equally indispensable, equally unsuppliable? They possess not, either in mind or body, the smallest thing exactly as thou dost. Each is wholly, and in every part as individual as thou art. Consider each as if he were single in the universe; then wilt thou discover powers and excellences in him which, abstractedly of comparision, deserve all attention and admiration. Compare him afterwards with others, his similarity, his dissimilarity to so many of his fellow-creatures. How must this incite thy amazement! How wilt thou value the individuality, the indispensability of his being! How wilt thou wonder at the harmony of his parts, each contributing to form one whole; at their relation, the relation of his million-fold individuality, to such multitudes of other individuals! Yes, we wonder at and adore the so simple, yet so infinitely varied expression of Almighty power inconceivable, so especially and so gloriously revealed in the nature of man.

No man ceases to be a man, how low soever he may sink beneath the dignity of human nature. Not being beast, he still is capable of amendment, of approaching perfection. The worst of faces still is a human face. Humanity ever continues the honour and ornament of man.

It is as impossible for a brute animal to become man, although he may in many actions approach, or almost surpass him, as for man to become a brute; although many men indulge themselves in actions which we cannot view in brutes without abhorrence.

But the very capacity of voluntarily debasing himself in appearance even below brutality, is the honour and privilege of man. This very capacity of imitating all things by an act of his will and the powers of his understanding, this very capacity man only has, beasts have not. The countenances of beasts are not susceptible of any remarkable deterioration, nor are they capable of any remarkable amelioration or beautifying. The worst of the countenances of men may be still more debased; but they may also, to a certain degree, be improved and ennobled.

The degree of perfection or degradation of which man is capable, cannot be described. For this reason the worst countenance has a well-founded claim to the notice, esteem, and hope of all good men. Again, in every human countenance, however debased, humanity is still visible; that is, the image of the Deity.

I have seen the worst of men in their worst of moments; yet could not all their vice, blasphemy, and oppression of guilt, extinguish the light of good that shone in their countenances, the spirit of humanity, the ineffaceable traits of internal, external perfectibility. The sinner we would exterminate, the man we must embrace. O, physiognomy, what a pledge art thou of the everlasting clemency of God towards man! O, man, rejoice with whatever rejoices in its existence, and condemn no being whom God doth not condemn!

CHAPTER VII.

Of the Forehead.

I SHALL appropriate this and some of the following chapters to remarks on certain individual parts of the human body. The following are my own remarks on foreheads :—

The form, height, arching, proportion, obliquity, and position of the skull or bone of the forehead, show the propensity, degree of power, thought, and sensibility of man; the covering or skin of the forehead, its position, colour, wrinkles, and tension, denote the passions and present state of the mind. The bones give the internal quantity, and their covering the application of power.

Though the skin be wrinkled, the forehead bones remain unaltered; but this wrinking varies according to the various forms of the bones. A certain degree of flatness produces certain wrinkles; a certain arching is attended by certain other wrinkles; so that the wrinkles, separately considered will give the arching; and this, *vice versâ*, will give the wrinkles. Certain foreheads can only have perpendicular, others horizontal, others curved, and others mixed and confused, wrinkles. Cup-formed (smooth) cornerless foreheads, when they are in motion, commonly have the simplest and least per-plexed wrinkles.

I consider the peculiar delineation of the outline and position of the forehead, which has been left unattempted by ancient and modern physiognomists, to be the most important of all the things presented to physiognomical observation. We may divide foreheads, considered in profile, into three principal classes, the retreating, the perpendicular, and the projecting. Each of these classes has a multitude of variations, which may easily again be classed, and the chief of which are rectilinear; half round, half rectilinear, flowing into each other; half round, half rectilinear, interrupted; curve-lined, simple; the curve-lined, double and triple.

I shall add some more particular remarks, which I apprehend will not be unacceptable to my readers:—

1. The longer the forehead, the more comprehension and less activity.

2. The more compressed, short, and firm the forehead, the more compression, firmness, and less volatility in the man.

3. The more curved and cornerless the outline, the more tender and flexible the character; the more rectili-near, the more pertinacity and severity.

4. Perfect perpendicularity, from the hair to the eye-brows, want of understanding.

5. Perfect perpendicularity, gently arched at the top, denotes excellent propensities of cold, tranquil, profound thinking.

6. Projecting—imbecility, immaturity, weakness, stupidity.

7. Retreating, in general, denotes superiority of imagination, wit, and acuteness.

8. The round and prominent forehead above, straight lined below, and on the whole perpendicular, shows much understanding, life, sensibility, ardour, and icy coldness.

9. The oblique, rectilinear forehead, is also very ardent and vigorous.

10. Arched foreheads appear properly to be feminine.

11. A happy union of straight and curved lines, with a happy position of the forehead, express the most perfect character of wisdom. By happy union, I mean when the lines insensibly flow into each other; and by happy position, when the forehead is neither too perpendicular nor too retreating.

12. I might almost establish it as an axiom, that right lines considered as such, and curves considered as such, are related, as power and weakness, obstinacy and flexibility, understanding and sensation.

13. I have hitherto seen no man with sharp projecting eye-bones who had not great propensity to an acute exercise of the understanding, and to wise plans.

14. Yet there are many excellent heads which have not this sharpness, and which have the more solidity, if the forehead, like a perpendicular wall, sink upon the horizontal eyebrows, and be greatly rounded on each side towards the temples.

15. Perpendicular foreheads, projecting so as not immediately to rest upon the nose, which are small, wrinkly, short, and shining, are certain signs of weakness, little understanding, little imagination, little sensation.

16. Foreheads with many angular, knotty protube-rances, ever denote much vigorous, firm, harsh, oppressive, ardent activity, and perseverance.

17. It is a sure sign of a clear, sound understanding, and a good temperament, when the profile of the forehead has two proportionate arches, the lower of which projects.

18. Eye-bones with defined, marked, easily delineated, firm arches, I never saw but in noble and in great men. All the ideal antiques have these arches.

19. Square foreheads, that is to say, with extensive temples and firm eye-bones, show circumspection and certainty of character.

20. Perpendicular wrinkles, if natural to the forehead, denote application and power; horizontal wrinkles, and those broken in the middle or at the extremities, in general, negligence or want of power.

21. Perpendicular deep indentings in the bones of the forehead, between the eyebrows, I never met with but in men of sound understanding, and free and noble minds, unless there were some positively contradictory feature.

22. A blue vena frontalis in the form of a Y, when in an open, smooth, well-arched forehead, I have only found in men of extraordinary talents, and of an ardent and generous character.

23. The following are the most indubitable signs of an excellent, a perfectly beautiful and significant, intel-ligent, and noble forehead :—

An exact proportion to the other parts of the counte-nance. It must equal the nose or the under part of the face in length, that is, one-third.

In breadth, it must either be oval at the top (like the foreheads of most of the great men of England) or nearly square.

A freedom from unevenness and wrinkles, yet with the power of wrinkling when deep in thought, afflicted by pain, or from just indignation.

Above it must retreat, project beneath.

The eye-bones must be simple, horizontal, and, if seen from above, must present a pure curve.

There should be a small cavity in the centre from above to below, and traversing the forehead so as to separate into four divisions, which can only be perceptible by a clear descending light.

The skin must be more clear in the forehead than in the other parts of the countenance.

The forehead must every where be composed of such outlines as, if the section of one-third only be viewed, it can scarcely be determined whether the lines are straight or circular.

24. Short, wrinkled, knotty, regular, pressed in one side, and sawcut foreheads, with intersecting wrinkles, are incapable of durable friendship.

25. Be not discouraged so long as a friend, an enemy, a child, or a brother, though a transgressor, has a good, well-proportioned, open forehead; there is still much certainty of improvement, much cause of hope.

CHAPTER VIII.

Of the Eyes and Eyebrows.

BLUE eyes are generally more significant of weakness, effeminacy, and yielding, than brown and black. True it is there are many powerful men with blue eyes; but I find more strength, manhood, and thought, combined with brown than with blue. Wherefore does it happen that the Chinese, or the people of the Philippine Islands, are very seldom blue-eyed; and that Europeans only, or the descendants of Europeans, have blue eyes in those countries? This is the more worthy of inquiry because there are no people more effeminate, luxurious, peaceable, or indolent than the Chinese.

Choleric men ·have eyes of every colour, but more brown, and inclined to green, than blue. This propensity to green is almost ·a decisive token of ardour, fire, and courage.

I have never met with clear blue eyes in the melancholic, seldom in the choleric; but most in the phlegmatic temperament, which, however, had much activity.

When the under arch described by the upper eyelid is perfectly circular, it always denotes goodness and tenderness, but also fear, timidity, and weakness.

The open eye, not compressed, forming a long acute angle with the nose, I have but seldom seen except in acute and understanding persons.

Hitherto I have seen no eye where the eyelid formed a horizontal line over the pupil, that did not appertain to a very acute, able, subtle man; but be it understood that I have met with this eye in very worthy men, but men of great penetration and simulation.

Wide, open eyes, with the white seen under the apple, I have often observed in the timid and phlegmatic, and also in the courageous and rash. When compared, however, the fiery and the feeble, the determined and the undetermined, will easily be distinguished. The former are more firm, more strongly delineated, have less obliquity, have thicker, better cut, but less skinny eyelids.

ADDITION.

From the Gotha Court Calendar, 1771, or rather from Buffon.

"The colours most common to the eyes are, the orange, yellow, green, blue, grey, and grey mixed with white. The blue and orange are most predominant, and are often found in the same eye. Eyes supposed to be black are only yellow, brown, or a deep orange; to convince ourselves of which, we need but look at them closely; for when seen at a distance, or turned towards the light, they appear to be black, because the yellow-brown colour is so contrasted to the white of the eye, that the opposition makes it supposed black. Eyes also of a less dark colour pass for black eyes, but are not esteemed so fine as the other, because the contrast is not so great. There are also yellow and light yellow eyes, which do not appear black, because the colours are not deep enough to be overpowered by the shade.

"It is not uncommon to perceive shades of orange, yellow, grey, and blue, in the same eye; and whenever blue appears, however small the tincture, it becomes the predominant colour, and appears in streaks over the whole iris. The orange is in flakes, round, and at some little distance from the pupil; but it is so strongly effaced by the blue that the eye appears wholly blue, and the mixture of orange is only perceived when closely inspected.

"The finest eyes are those which we imagine to be black or blue. Vivacity and fire, which are the principal characteristics of the eyes, are the more emitted when the colours are deep and contrasted, rather than when slightly shaded. Black eyes have most strength of expression, and most vivacity; but the blue have most mildness, and perhaps are more arch. In the former there is an ardour uninterruptedly bright, because the colour, which appears to us uniform, every way emits similar reflections. But modifications are distinguished in the light which animates blue eyes, because there are various tints of colour which produce various reflections.

"There are eyes which are remarkable for having what may be said to be no colour. They appear to be differently constituted from others. The iris has only some shades of blue or grey, so feeble that they are in some parts almost white; and the shades of orange which intervene are so small that they scarcely can be distinguished from grey or white, notwithstanding the contrast of these colours. The black of the pupil is then too marking, because the colour of the iris is not deep enough, and, as I may say, we see only the pupil in the centre of the eye. These eyes are unmeaning, and appear to be fixed and aghast.

"There are also eyes the colour of the iris of which is almost green; but these are more uncommon than the blue, the grey, the yellow, and the yellow-brown. There are also people whose eyes are not both of the same colour.

"The images of our secret agitations are particularly painted in the eyes. The eye appertains more to the soul than any other organ; seems affected by, and to participate in, all its motions; expresses sensations the most lively, passions the most tumultuous, feelings the most delightful, and sentiments the most delicate. It explains them in all their force, in all their purity, as they take birth; and transmits them by traits so rapid as to infuse into other minds the fire, the activity, the very image with which themselves are inspired. The eye at once receives and reflects the intelligence of thought, and the warmth of sensibility. It is the sense of the mind, the tongue of the understanding."

Again, "As in nature, so in art, the eyes are differently formed in the statues of the gods, and in heads of ideal beauty; so that the eye itself is the distinguishing token. Jupiter, Juno, and Apollo, have large, round, well-arched eyes, shortened in length, in order that the arch may be the higher. Pallas, in like manner, has large eyes; but the upper eyelid, which is drawn up, is expressive of attraction and languishment. Such an eye distinguishes the heavenly Venus Urania from Juno; yet the statue of this Venus bearing a diadem, has for that reason often been mistaken, by those who have not made this observation, for the statue of Juno. Many of the modern artists appear to have been desirous of excelling the ancients, and to give what Homer calls the ox-eye, by making the pupil project, and seem to start from the socket. Such an eye has the modern head of the erroneously supposed Cleopatra, in the Medicean villa, and which presents the idea of a person strangled. The same kind of eye a young artist has given to the statue of the Holy Virgin, in the church St. Carlo al Torso."

I shall quote one more passage from Paracelsus, who, though an astrological enthusiast, was a man of prodigious genius :—

"To come to the practical part, and give proper signs with some of their significations, it is to be remarked that blackness in the eyes generally denotes health, a firm mind—not wavering, but courageous, true, and honourable. Grey eyes generally denote deceit, instability, and indecision. Short sight denotes an able projector, crafty, and intriguing in action. The squinting or false-sighted, who see on both sides, or over and under, certainly denotes a deceitful crafty person, not easily deceived, mistrustful, and not always to be trusted; one who willingly avoids labour when he can, indulging in idleness, play, usury, and pilfering. Small and deep sunken eyes are bold in opposition; not discouraged, intriguing, and active in wickedness; capable of suffering much. Large eyes denote a covetous greedy man, and especially when they are prominent. Eyes in continual motion signify short or weak sight, fear, and care. The winking eye denotes an amorous disposition and foresight, and quickness in projection. The downcast eye shows shame and modesty. Red eyes signify courage and strength. Bright eyes, slow of motion, bespeak the hero, great acts, audacious, cheerful, one feared by his enemies."

It will not be expected I should subscribe to all these opinions, they being most of them ill-founded, at least ill-defined.

The Eyebrows.

Eyebrows regularly arched are characteristic of feminine youth; rectilinear and horizontal, are masculine; arched and the horizontal combined, denote masculine understanding and feminine kindness.

Wild and complexed, denote a corresponding mind, unless the hair be soft, and they then signify gentle ardour.

Compressed, firm, with the hair running parallel as if cut, are one of the most decisive signs of a firm, manly, mature understanding, profound wisdom, and a true and unerring perception.

Meeting eyebrows, held so beautiful by the Arabs, and by the old physiognomists supposed to be the mark of craft, I can neither believe to be beautiful nor characteristic of such a quality. They are found in the most open, honest, and worthy countenances. It is true they give the face a gloomy appearance, and perhaps denote trouble of mind and heart.

Sunken eyebrows, says Winkelmann, impart something of the severe and melancholy to the head of Antinous.

I never yet saw a profound thinker, or even a man of fortitude and prudence, with weak, high eyebrows, which in some measure equally divide the forehead.

Weak eyebrows denote phlegm and debility, though there are choleric and powerful men who have them; but this weakness of eyebrows is always a deduction from power and ardour.

Angular, strong, interrupted eyebrows, ever denote fire and productive activity.

The nearer the eyebrows are to the eyes, the more earnest, deep, and firm the character.

The more remote from the eyes, the more volatile, easily moved, and less enterprising.

Remote from each other, warm, open, quick sensation.

White eyebrows signify weakness; and dark brown firmness.

The motion of the eyebrows contains numerous expressions, especially of ignoble passions, pride, anger, and contempt.

CHAPTER IX.

Of the Nose.

I HAVE generally considered the nose as the foundation or abutment of the brain. Whoever is acquainted with the Gothic arch, will perfectly understand what I mean by this abutment: for upon this the whole power of the arch of the forehead rests, and without it the mouth and cheeks would be oppressed by miserable ruins.

A beautiful nose will never be found accompanying an ugly countenance. An ugly person may have fine eyes, but not a handsome nose. I meet with thousands of beautiful eyes before one such nose; and wherever I find the latter it denotes an extraordinary character. The following is requisite to the perfectly beautiful nose :—

Its length should equal the length of the forehead. At the top should be a gentle indenting. Viewed in front, the back should be broad, and nearly parallel, yet above the centre something broader. The button or end of the nose must be neither hard nor fleshy, and its under outline must be remarkably definite, well delineated, neither pointed nor very broad. The sides seen in front must be well defined, and the descending nostrils gently shortened. Viewed in profile, the bottom of the nose should not have more than one-third of its length. The nostrils above must be pointed; below, round, and have in general a gentle curve, and be divided into two equal parts by the profile of the upper lip. The sides or arch of the nose must be a kind of wall. Above, it must close well with the arch of the eye-bone, and near the eye must be at least half an inch in breadth. Such a nose is of more worth than a kingdom. There are, indeed, innumerable excellent men with defective noses, but their excellence is of a very different kind. I have seen the purest, most capable, and noblest persons, with small noses, and hollow in profile; but their worth most consisted in suffering, listening, learning, and enjoying the beautiful influences of imagination; provided the other parts of the form were well organized. Noses, on the contrary, which are arched near the forehead, are capable of command, can rule, act, overcome, destroy. Rectilinear noses may be called the keystone between the two extremes. They equally act and suffer with power and tranquillity.

Boerhaave, Socrates, Lairesse, had, more or less, ugly noses, and yet were great men; but their character was that of gentleness and patience.

I have never yet seen a nose with a broad back, whether arched or rectilinear, that did not appertain to an extraordinary man. We may examine thousands of countenances, and numbers of portraits of superior men, before we find such a one.

These noses were possessed, more or less, by Raynal, Faustus Socinus, Swift, Cæsar Borgia, Clepzecker, Anthony Pagi, John Charles von Enkenberg (a man of Herculean strength), Paul Sarpi, Peter de Medicis, Francis Caracci, Casini, Lucas van Leyden, Titian.

There are also noses that are not broad backed, but small near the forehead, of extraordinary power; but their power is rather elastic and momentary than productive.

The Tartars generally have flat indented noses; the negroes broad, and the Jews hawk noses. The noses of

Englishmen are seldom pointed, but generally round. The Dutch, if we may judge from their portraits, seldom have handsome or significant noses. The nose of the Italian is large and energetic. The great men of France, in my opinion, have the characteristic of their greatness generally in the nose : to prove which, examine the collection of portraits by Perrault and Morin.

Small nostrils are usually an indubitable sign of unenterprising timidity. The open breathing nostril is as certain a token of sensibility, which may easily degenerate into sensuality.

CHAPTER X.

Of the Mouth and Lips.

THE contents of the mind are communicated to the mouth. How full of character is the mouth, whether at rest or speaking, by its infinite powers !

Whoever internally feels the worth of this member, so different from every other member, so inseparable, so not to be defined, so simple, yet so various ; whoever, I say, knows and feels this worth, will speak and act with divine wisdom. Oh ! wherefore can I only imperfectly and tremblingly declare all the honours of the mouth— the chief seat of wisdom and folly, power and debility, virtue and vice, beauty and deformity, of the human mind—the seat of all love, all hatred, all sincerity, all falsehood, all humility, all pride, all dissimulation, and all truth ?

Oh ! with what adoration would I speak, and be silent, were I a more perfect man ! Oh ! discordant, degraded humanity ! Oh ! mournful secret of my misinformed

youth! When, Omniscience, shalt thou stand revealed?
Unworthy as I am, yet do I adore. Yet worthy I shall
be; worthy as the nature of man will permit: for he
who created me gave a mouth to glorify Him!

Painters and designers, what shall I say that may induce you to study this sacred organ, in all its beauteous
expressions, all its harmony and proportion?

Take plaster impressions of characteristic mouths of
the living and the dead; draw after, pore over them;
learn, observe, continue day after day to study one only;
and, having perfectly studied that, you will have studied
many. Oh! pardon me; my heart is oppressed. Among
ten or twenty draughtsmen, to whom for three years I
have preached, whom I have instructed, have drawn examples for, not one have I found who felt as he ought to
feel, saw what was to be seen, or could represent that
which was evident. What can I hope?

Every thing may be expected from a collection of
characteristic plaster impressions, which might so easily
be made were such a collection only once formed. But
who can say whether such observations might not declare
too much? The human machine may be incapable of
suffering to be thus analyzed. Man, perhaps, might not
endure such close inspection; and therefore, having eyes,
he sees not. I speak it with tears; and why I weep,
thou knowest, who with me inquirest into the worth of
man. And you weaker yet candid, though on this occasion unfeeling, readers, pardon me!

Observe the following rules: Distinguish in each
mouth the upper lip singly; the under lip the same;
the line formed by the union of both when tranquilly
closed, if they can be closed without constraint; the
middle of the upper lip, in particular, and of the under

lip; the bottom of the middle line at each end; and, lastly, the extending of the middle line on both sides. For, unless you thus distinguish, you will not be able to delineate the mouth accurately.

As are the lips, so is the character. Firm lips, firm character; weak lips and quick in motion, weak and wavering character.

Well defined, large, and proportionate lips, the middle line of which is equally serpentine on both sides, and easy to be drawn, though they may denote an inclination to pleasure, are never seen in a bad, mean, common, false, crouching, vicious countenance.

A lipless mouth, resembling a single line, denotes coldness, industry, a love of order, precision, housewifery; and, if it be drawn upwards at the two ends, affectation, pretension, vanity, and, which may ever be the production of cool vanity, malice.

Very fleshy lips must ever have to contend with sensuality and indolence: the cut-through, sharp-drawn lip, with anxiety and avarice.

Calm lips, well closed without constraint, and well delineated, certainly betoken consideration, discretion, and firmness.

A mild overhanging upper lip generally signifies goodness. There are innumerable good persons also with projecting under lips; but the goodness of the latter is rather cold fidelity and well-meaning, than warm active friendship.

The under lip, hollowed in the middle, denotes a fanciful character. Let the moment be remarked when the conceit of the jocular man descends to the lip, and it will be seen to be a little hollow in the middle.

A closed mouth, not sharpened, not affected, always

denotes courage and fortitude; and the open mouth always closes where courage is indispensable. Openness of mouth speaks complaint; and closeness, endurance.

Though physiognomists have as yet but little noticed, yet much might be said concerning the lip improper, or the fleshy covering of the upper teeth, on which anatomists have not, to my knowledge, yet bestowed any name, and which may be called the curtain, or pallium, extending from the beginning of the nose to the red upper lip proper.

If the upper lip improper be long, the proper is always short; if it be short and hollow, the proper will be large and curved—another certain demonstration of the conformity of the human countenance. Hollow upper lips are much less common than flat and perpendicular; the character they denote is equally uncommon.

CHAPTER XI.

Of the Teeth and Chin.

NOTHING is more striking, or continually visible, than the characteristics of the teeth, and the manner in which they display themselves. The following are the observations I have made thereon :—

Small short teeth, which have generally been held by the old physiognomists to denote weakness, I have remarked in adults of extraordinary strength; but they seldom were of a pure white.

Long teeth are certain signs of weakness and pusilla nimity. White, clean, well-arranged teeth, visible as soon as the mouth opens, but not projecting, nor always entirely seen, I have never met with in adults, except in good, acute, honest, candid, faithful men.

I have also met foul, uneven, and ugly teeth, in persons of the above good character; but it was always either sickness, or some mental imperfection, which gave this deformity.

Whoever leaves his teeth foul, and does not attempt to clean them, certainly betrays much of the negligence of his character, which does him no honour.

As are the teeth of man, that is to say, their form, position, and cleanliness, (so far as the latter depends on himself,) so is his taste.

Wherever the upper gum is very visible at the first opening of the lips, there is generally much cold and phlegm.

Much, indeed, might be written upon the teeth, though they are generally neglected in all historical paintings. To be convinced of this, we need but observe the teeth of an individual during the course of a single day, or contemplate an apartment crowded with fools. We should not then, for a moment, deny that the teeth, in conjunction with the lips, are very charactcristic; or that physiognomy has gained another token which triumphs over all the arts of dissimulation.

The Chin.

I am, from numerous experiments, convinced that the projecting chin ever denotes something positive, and the retreating something negative. The presence or absence of strength in man is often signified by the chin.

I have never seen sharp indentings in the middle of the chin but in men of cool understanding, unless when something evidently contradictory appeared in the countenance.

The pointed chin is generally held to be a sign of acuteness and craft, though I know very worthy persons with such chins. Their craft is the craft of the best dramatic poetry.

The soft, fat, double chin generally points out the epicure; and the angular chin is seldom found but in discreet, well-disposed, firm men.

Flatness of chin speaks the cold and dry; smallness, fear; and roundness, with a dimple, benevolence.

CHAPTER XII.

Of Skulls.

How much may the anatomist see in the mere skull of man! How much more the physiognomist! And how much the most the anatomist who is a physiognomist! I blush when I think how much I ought to know, and of how much I am ignorant, while writing on a part of the body of man which is so superior to all that science has yet discovered—to all belief, to all conception!

I consider the system of the bones as the great outline of man, and the skull as the principal part of that system. I pay more attention to the form and arching of the skull, as far as I am acquainted with it, than all my predecessors; and I have considered this most firm, least changeable, and far best defined part of the human body, as the foundation of the science of physiognomy. I shall therefore be permitted to be particular in my observations on this member of the human body.

I confess, that I scarcely know where to begin, where to end, what to say, or what to omit. I think it advisable to premise a few words concerning the generation and formation of human bones.

The whole of the human fœtus is at first supposed to be only a soft mucilaginous substance, homogeneous in all its parts; and that the bones themselves are but a kind of coagulated fluid, which afterwards becomes membraneous, then cartilaginous, and at last hard bone.

As this viscous congelation, originally so transparent and tender, increases, it becomes thicker and more opaque, and a dark point makes its appearance different from the cartilage, and of the nature of bone, but not yet perfectly hard. This point may be called the kernel of the future bone, the centre round which the ossification extends.

We must, however, consider the coagulation attached to the cartilage as a mass without shape, and only with a proper propensity for assuming its future form. In its earliest, tenderest state, the traces of it are expressed upon the cartilage, though very imperfectly.

With respect to the bony kernels, we find differences which seem to determine the form of the future bones. The simple and smaller bones have each only one kernel; but in the more gross, thick, and angular, there are several in different parts of the original cartilage; and it must be remarked that the number of the joining bones is equivalent to the number of the kernels.

In the bones of the skull the round kernel first is apparent in the centre of each piece; and the ossification extends itself, like radii from the centre, in filaments, which increase in length, thickness, and solidity, and are interwoven with each other like network. Hence these delicate, indented features of the skull, when its various parts are at length joined.

We have hitherto only spoken of the first stage of ossification. The second begins about the fourth or fifth month, when the bones, together with the rest of the parts, are more perfectly formed, and, in the progress of ossification, include the whole cartilage, according to the more or less life of the creature, and the original different impulse and power of motion in the being.

Agreeable to their original formation through each succeeding period of age, they will continue to increase in thickness and hardness. But on this subject anatomists disagree—so let them. Future physiognomists may consider this more at large. I retreat from contest, and will travel in the high-road of certainty, and confine myself to what is visible.

Thus much is certain, that the activity of the muscles, vessels, and other parts which surround the bones, contributes much to their formation, and gradual increase in hardness.

The remains of the cartilaginous in the young bones will, in the sixth and seventh month, decrease in quantity, harden, and whiten, as the bony parts approach perfection. Some bones obtain a certain degree of firmness in much less time than others; as, for example, the skull bones, and the small bones within the ear. Not only whole bones, but parts of a single bone, are of various degrees of hardness. They will be hardest at the place where the kernel of ossification began, and the parts adjacent; and the rigidity increases more slowly and insensibly the harder the bones are, and the older the man is. What was cartilage will become bone; parts that were separate will grow together, and the whole bones be deprived of moisture.

Anatomists divide the form into the natural or the

essential, which is generally the same in all bones in the human body, how different soever it may be to other bodies; and into the accidental, which is subject to various changes in the same individual, according to the influence of external objects, or, especially, of the gradations of age.

The first is founded in the universality of the nature of parents, and the circumstances which naturally and invariably attend propagation. Anatomists consider only the designation of the bones individually; on this, at least, is grounded the agreement of what they call the essential form in distinct subjects. This, therefore, only speaks to the agreement of human countenances, so far as they have each two eyes, one nose, one mouth, and other features thus or thus disposed.

This natural formation is certainly as different as human countenances afterwards are; which difference is the work of nature, the original destination of the Lord and Creator of all things. The physiognomist distinguishes between original form and deviations.

Each bone hath its original form, its individual capacity of form. It may, it does continually alter; but it never acquires the peculiar form of another bone, which was originally different. The accidental changes of bones, however great, or different from the original form, are yet ever governed by the nature of this original individual form; nor can any power of pressure ever so change the original form, but that, if compared to another system of bones that has suffered an equal pressure, it will be perfectly distinct. As little as the Ethiopian can change his skin, or the leopard his spots, whatever be the changes to which they may be subject, as little can the original form of any bone be changed into the original form of any other bone.

Vessels every where penetrate the bones, supplying them with juices and marrow. The younger the bone is the more are there of these vessels; consequently the more porous and flexible are the bones, and the reverse. The period when such or such changes take place in the bones cannot easily be defined; it differs according to the nature of men in accidental circumstances.

Large and long and multiform bones, in order to facilitate their ossification and growth, at first consist of several pieces, the smaller of which are called supplemental. The bone remains imperfect till these become incorporated. Hence their possible distortion in children by the rickets, and other diseases.

CHAPTER XIII.

Suggestions to the Physiognomist concerning the Skull.

THE scientific physiognomist ought to direct his attention to the distortion of the bones, especially those of the head. He ought to learn accurately to remark, compare, and define, the first form of children, and the numerous relative deviations. He ought to have attained that precision that should enable him to say, at beholding the head of a new-born infant of half a year, a year, or two years old, "Such and such will be the form of the system of the bones, under such and such limitations;" and on viewing the skull at ten, twelve, twenty, or twenty-four years of age, "Such or such was the form, eight, ten, or twenty years ago; and such or such will be the form, eight, ten, or twenty years hence, violence excepted." He ought to be able to see the youth in the boy, and the man in the youth; and, on the reverse, the youth in the man, the boy in the youth, the infant in the boy, and, lastly, the embryo in its proper individual form.

Let us, O ye who adore that Wisdom which has framed all things! contemplate a moment longer the human skull. There are, in the bare skull of man, the same varieties as are to be found in the whole external form of the living man.

As the infinite variety of the external form of man is one of the indestructible pillars of physiognomy, no less so, in my opinion, must the infinite varieties of the skull itself be. What I have hereafter to remark will, in part, show that we ought particularly to begin by that, if, instead of a subject of curiosity and amusement, we would wish to make the science of physiognomy universally useful.

I shall show that from the structure, form, outline, and properties of the bones, not all indeed, but much may be discovered, and probably more than from all the other parts.

Objection and Answer.

What answer shall I make to that objection, with which a certain anti-physiognomist has made himself so merry?

"In the catacombs near Rome," says he, "a number of skeletons were found, which were supposed to be the relics of saints, and as such were honoured. After some time, several learned men began to doubt whether these had really been the sepulchres of the first Christians and martyrs, and even to suspect that malefactors and banditti might have been buried there. The piety of the faithful was thus much puzzled; but if the science of physiognomy be so certain, they might have removed all their doubts by sending for Lavater, who with very little trouble, by merely examining and touching them, might have distinguished the bones of the saints from the bones of the banditti, aud thus have restored the true relics to their just and original pre-eminence."

"The conceit is whimsical enough," answers a cold and phlegmatic friend of physiognomy; "but, having tired ourselves with laughing, let us examine what would have been the consequence had this story been fact. According to our opinion, the physiognomist would have remarked great differences in a number of bones, particularly in the skulls, which to the ignorant would have appeared perfectly similar; and having classed his heads, and shown their immediate gradations, and the contrast of the two extremes, we may presume the attentive spectator would have been inclined to pay some respect to his conjectures on the qualities and activity of brain which each formerly contained.

"Besides, when we reflect how certain it is that many malefactors have been possessed of extraordinary abilities and energy, and how uncertain it is whether many of the saints who are honoured with red-letter days in the calendar ever possessed such qualities, we find the question so intricate that we should be inclined to pardon the poor physiognomist were he to refuse an answer, and leave the decision to the great infallible Judge."

Further Reply.

Let us endeavour further to investigate the question; for, though this answer is good, it is insufficient. Who

ever yet pretended absolutely to distinguish saints from
banditti, by inspecting only the skull.

To me it appears that justice requires we should, in
all our decisions concerning books, men, and opinions,
judge each according to their pretensions, and not ascribe
pretensions which have not been made to any man.

I have heard of no physiognomist who has had, and I
am certain that I myself never have had, any such presumption. Notwithstanding which, I maintain as a
truth most demonstrable, that by the mere form, proportion, hardness, or weakness of the skull, the strength
or weakness of the general character may be known with
the greatest certainty. But, as hath been often repeated,
strength and weakness are neither virtue nor vice, saint
nor malefactor.

Power, like riches, may be employed to the advantage
or detriment of society, the same as wealth may be in
the possession of a saint or a demon; and as it is with
wealth or arbitrary positive power, so is it with natural
innate power. As in an hundred rich men there are
ninety-nine who are not saints, so will there scarcely
be one saint among an hundred men born with this
power.

When, therefore, we remark in a skull great original
and percussive power, we cannot indeed say this man
was a malefactor; but we may affirm there was this
excess of power which, if it were not qualified and tem-
pered during life, there is the highest probability it
would have been agitated by the spirit of conquest,
would have become a general, a conqueror, a Cæsar, a
Cartouch. Under certain circumstances he would probably have acted in a certain manner, and his actions
would have varied according to the variation of circum-

stances; but he would always have acted with ardour, tempestuously—always as a ruler and a conqueror.

Thus, also, we may affirm of certain other skulls which in their whole structure and form discover ten derness, and resemblance to parchment, that they denote weakness; a mere capability of perceptive without percussive, without creative power. Therefore, under certain circumstances, such persons would have acted weakly. They would not have had the native power of withstanding this or that temptation, of engaging in this or that enterprise. In the fashionable world they would have acted the fop, the libertine in a more confined circle, and the enthusiastic saint in a convent.

Oh! how differently may the same power, the same sensibility, the same capacity, act, feel, and conceive under different circumstances! And hence we may, in part, comprehend the possibility of predestination and liberty in one and the same subject.

Take a man of the commonest understanding to a charnel-house, and make him attentive to the differences of skulls; in a short time he will either perceive of himself, or understand when told, here is strength, there weakness; here obstinacy, and there indecision.

If shown the bald head of Cæsar as painted by Rubens or Titian, or that of Michael Angelo, what man would be dull enough not to discover that impulsive power, that rocky comprehension, by which they were peculiarly characterised; and that more ardour, more action, must be expected than from a smooth, round, flat head?

How characteristic is the skull of Charles XII.! How different from the skull of his biographer, Voltaire! Compare the skull of Judas with the skull of Christ after Holbein, discarding the muscular parts, and I doubt, if asked which was the wicked betrayer, which the innocent betrayed, whether any one would hesitate.

I will acknowledge that, when two determinate heads are presented to us with such striking differences, the one of which is known to be that of a malefactor, the other that of a saint, it is infinitely more easy to decide; but he who can distinguish between them, should not therefore affirm he can distinguish the skulls of saints from the skulls of malefactors.

To conclude this chapter. Who is unacquainted with the anecdote of Herodotus, that it was possible many years afterwards, on the field of battle, to distinguish the skulls of the effeminate Medes from those of the manly Persians? I think I have heard the same remark made of the Swiss and the Burgundians. This at least proves it is granted that we may perceive, in the skull only, a difference of strength and manners as well as of nations.

CHAPTER XIV.

Of the Difference of Skulls as they relate to Sex, and particularly to Nations.—Of the Skulls of Children.

An Essay on the difference of bones, as they relat to sex, and particularly to nations, has been published by M. Fischer, which is well deserving of attention. The following are some thoughts on the subject, concerning which nothing will be expected from me, but very much from M. Kamper.

Consideration and comparison of the external and internal make of the body, in male and female, teaches us that the one is destined for labour and strength, and the other for beauty and propagation. The bones particularly denote masculine strength in the former; and, so far as the stronger and the prominent are more easy to describe than the less prominent and the weaker, so far is the male skeleton and the skull the easier to define.

The general structure of the bones in the male, and of the skull in particular, is evidently of stronger formation than in the female. The body of the male increases, from the hip to the shoulder, in breath and thickness; hence the broad shoulders and square form of the strong: whereas the female skeleton gradually grows thinner and weaker from the hip upwards, and by degrees appears as if it were rounded.

Even single bones in the female are more tender, smooth, and round ; have fewer sharp edges, cutting and prominent corners.

We may here properly cite the remark of Santorinus concerning the difference of skulls as they relate to sex. "The aperture of the mouth, the palate, and in general the parts which form the voice, are less in the female; and the more small and round chin, consequently the under part of the mouth, correspond."

The round or angular form of the skull may be very powerfully and essentially turned to the advantage of the physiognomist, and becomes a source of innumerable individual judgments. Of this the whole work abounds with proofs and examples.

No man is perfectly like another, either in external construction or internal parts, whether great or small, or in the system of the bones. I find this difference not only between nations, but between persons of the nearest kindred; but not so great between these, and between persons of the same nation, as between nations remote from each other, whose manners and food are very different. The more confidently men converse with, the more they resemble each other, as well in the formation of the parts of the body, as in language, manners, and food; that is, so far as the formation of the body can be influenced by external accidents. Those nations, in a certain degree, will resemble each other that have commercial intercourse, they being acted upon by the effect of climate, imitation, and habit, which have so great an influence in forming the body and mind—that is to say, the visible and invisible powers of man; although national character still remains, and which character, in reality, is much easier to remark than to describe.

We shall leave more extensive inquiries and observations concerning this subject to some such person as Kamper, and refrain, as becomes us; not having obtained sufficient knowledge of the subject to make remarks of our own of sufficient importance.

Differences with respect to strength, firmness, structure, and proportion of the parts, are certainly visible in all the bones of the skeletons of the different nations; but most in the formation of the countenance, which every where contains the peculiar expression of nature, of the mind.

The skull of a Dutchman, for example, is in general rounder, with broader bones, curved, and arched in all its parts, and with the sides less flat and compressed.

A Calmuc skull will be more rude and gross; flat on the top, prominent at the sides; the parts firm and compressed, the face broad and flat.

The skull of the Ethiopian is steep, suddenly elevated;

as suddenly small, sharp above the eyes; beneath strongly projecting; circular and high behind.

In proportion as the forehead of the Calmuc is flat and low, that of the Ethiopian is high and narrow; while the back part of an European head has a much more protuberant arch, and spherical form behind, than that of a negro.

Of the Skulls of Children.

The skull or head of a child, drawn upon paper, without additional circumstance, will be generally known, and seldom confounded with the head of an adult. But, to keep them distinct, it is necessary the painter should not be too hasty and incorrect in his observations of what is peculiar, or so frequently generalize the particular, which is the eternal error of painters, and of so many pretended physiognomists.

Notwithstanding individual variety, there are certain constant signs proper to the head of a child, which as much consist in the combination and form of the whole, as in the single parts.

It is well known that the head is larger in proportion to the rest of the body, the younger the person is; and it seems to me, from comparing the skull of the embryo, the child, and the man, that the part of the skull which contains the brain is proportionably larger than the parts that compose the jaw and the countenance. Hence it happens that the forehead in children, especially the upper part, is generally so prominent.

The bones of the upper and under jaw, with the teeth they contain, are later in their growth, and more slowly attain perfect formation. The under part of the head generally increases more than the upper, till it has attained full growth. Several processes of the bones, as the *processus mamillares*, which lie behind and under the ears, form themselves after the birth; as do also, in a great measure, various hidden sinuses or cavities in these bones. The quill form of these bones, with their various points, ends, and protuberances, and the numerous muscles which are annexed to them, and continually in action, make the greater increase and change more possible and easy than can happen in the spherical bony covering of the brain, when once the sutures are entirely become solid.

This unequal growth of the two principal parts of the skull must necessarily produce an essential difference in the whole, without enumerating the obtuse extremities, the edges, sharp corners, and single protuberances, which are chiefly occasioned by the action of the muscles.

As the man grows, the countenance below the forehead becomes more protuberant; and as the sides of the face, that is to say, the temple bones, which are also slow in coming to perfection, continually remove further from each other, the skull gradually loses that pear form which it appears to me to have had in embryo.

The *sinus frontales* first form themselves after birth. The prominence at the bottom of the forehead, between the eyebrows, is likewise wanting in children. The forehead joins the nose without any remarkable curve. This latter circumstance may also be observed in some grown persons, when the *sinus frontales* are either wanting or very small; for these cavities are found very different in different subjects.

The nose, during growth, alters exceedingly; but I am unable to explain in what manner the bones contribute to this alteration, it being chiefly cartilaginous. Accurately to determine this, many experiments on the heads and skulls of children, and grown persons, would be necessary; or, rather, if we could compare the same head with itself at different ages, which might be done by the means of shades, such gradation of the head or heads would be of great utility to the physiognomist.

CHAPTER XV.

Description of Plate III.

Number 1.

THIS outline, from a bust of Cicero, appears to me an almost perfect model of congeniality; the whole has the character of penetrating acuteness, an extraordinary though not a great profile. All is acute; all is sharp: discerning, searching, less benevolent than satirical, elegant, conspicuous, subtle.

Number 2.

Another congenial countenance. Too evidently nature for it to be mistaken for ideal, or the invention and emendation of art. Such a forehead does not betoken the rectilinear, but the nose thus bent. Such an upper lip, such an open, eloquent mouth! The forehead does not lead us to expect high poetical genius; but acute punctuality, and the stability of retentive memory. It is impossible to suppose this a common countenance.

Number 3.

The forehead and nose not congenial. The nose shows the very acute thinker. The lower part of the forehead,

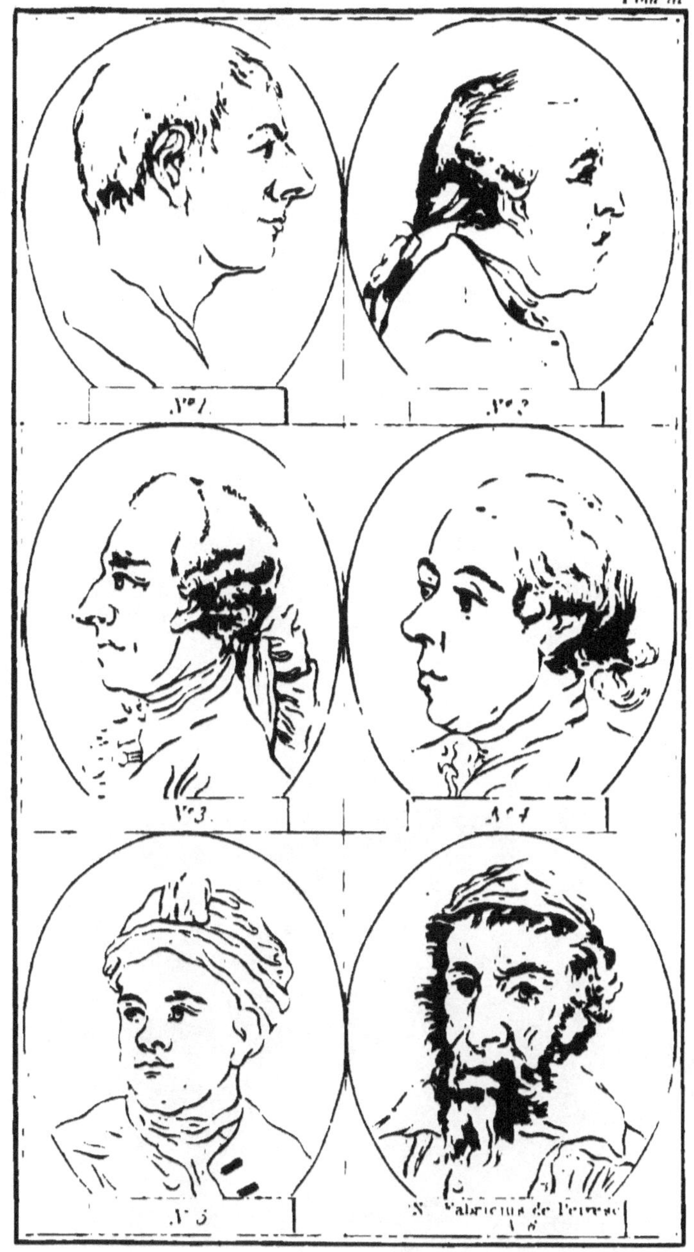
Nº 1.
Nº 2.
Nº 3.
Nº 4.
Nº 5.
N. Fabricius de Peiresc
Nº 6

on the contrary, especially the distance between the eyebrow and eye, do not betoken this high degree of mental power. The stiff position of the whole is much at variance with the eye and mouth, but particularly with the nose. The whole, the eyebrow excepted, speaks a calm, peaceable, mild character.

Number 4.

The harmony of the mouth and nose is self-evident. The forehead is too good, too comprehensive, for this very limited under part of the countenance. The whole bespeaks a harmless character; nothing delicate nor severe.

Number 5.

We have here a high bold forehead, with a short-seeming blunt nose, and a fat double chin. How do these harmonize! It is almost a general law of nature, that where the eyes are strong drawn, and the eyebrows near, the eyebrows must also be strong. This countenance, merely by its harmony, its prominent congenial traits, is expressive of sound, clear understanding; the countenance of reason.

Number 6.

The perfect countenance of a politician. Faces which are thus pointed from the eyes to the chin always have lengthened noses, and never possess large, open, powerful, and piercing eyes. Their firmness . partakes of obstinacy, and they rather follow intricate plans than the dictates of common-sense.

CHAPTER XVI.

The Physiognomist.

ALL men have talents for all things; yet we may venture to assert that very few have the determinate and essential talents. All men have talents for drawing: they can all learn to write, well or ill; yet not an excellent draftsman will be produced in ten thousand. The same may be affirmed of eloquence, poetry, and physiognomy. All men who have eyes and ears have talents to become physiognomists; yet not one in ten thousand can become an excellent physiognomist.

It may therefore be of use to sketch the character of the true physiognomist, that those who are deficient of the requisite talents may be deterred from the study of physiognomy. The pretended physiognomist, with a foolish head and a wicked heart, is certainly one of the most contemptible and mischievous creatures that crawls on the face of the earth.

No one whose person is not well formed can become a good physiognomist. Those painters were the best whose persons were the handsomest. Rubens, Vandyke, and Raphael, possessing three gradations of beauty, possessed three gradations of the genius of painting. The physiognomists of the greatest symmetry are the best. As the most virtuous can best determine on virtue, and the just on justice, so can the most handsome countenances on the goodness, beauty, and noble traits of the human countenance, and consequently on its defects and ignoble properties. The scarcity of human beauty is the reason why physiognomy is so much decried, and finds so many opponents.

No person, therefore, ought to enter the sanctuary of physiognomy who has a debased mind, an ill-formed forehead, a blinking eye, or a distorted mouth. "The light of the body is the eye; if, therefore, thine eye be single, thy whole body shall be full of light; but if thine eye be evil, thy whole body shall be full of darkness: if, therefore, the light that is in thee be darkness, how great is that darkness!"

Any one who would become a physiognomist cannot meditate too much on this text. O single eye! that beholdest all things as they are, seest nothing falsely, with glance oblique, nothing overlookest! O most perfect image of reason and wisdom!—why do I say image?—thou art reason and wisdom themselves! Without thy resplendent light would all that appertains to physiognomy become dark!

He who does not, at the first aspect of any man, feel a certain emotion of affection or dislike, attraction or repulsion, never can become a physiognomist.

He who studies art more than nature, and prefers what the painters call manner to the truth of drawing he who does not feel himself moved almost to tears, at beholding the ancient ideal beauty, and the present depravity of men and imitative art; he who views antique gems, and does not discover enlarged intelligence in Cicero, enterprising resolution in Cæsar, profound thought in Solon, invincible fortitude in Brutus, in Plato godlike wisdom; or, in modern medals, the height of human sagacity in Montesquieu, in Haller the energetic contemplative look, and the most refined taste; the deep reasoner in Locke, and the witty satirist in Voltaire, even at the first glance, never can become a physiognomist.

G

He who does not dwell with fixed rapture on the aspect of benevolence in action, supposing itself unobserved; he who remains unmoved by the voice of innocence, the guiltless look of unviolated chastity, the mother contemplating her beauteous sleeping infant; the warm pressure of the hand of a friend, or his eye swimming in tears; he who can lightly tear himself from scenes like these, and turn them to ridicule, might much easier commit the crime of parricide than become a physiognomist.

If such be the case, what then is required of the physiognomist? What should his inclination, talents, qualities, and capabilities be?

In the first place, as hath been in part already remarked, his first of requisites should be a body well proportioned and finely organized; accuracy of sensation, capable of receiving the most minute outward impressions, and easily transmitting them faithfully to memory, or, as I ought rather to say, impressing them upon the imagination and the fibres of the brain. His eye, in particular, must be excellent, clear, acute, rapid, and firm.

The very soul of physiognomy is precision in observation. The physiognomist must possess a most delicate, swift, certain, most extensive spirit of observation. To observe is to be attentive, so as to fix the mind on a particular object, which it selects, or may select, for consideration, from a number of surrounding objects. To be attentive is to consider some one particular object, exclusively of all others, and to analyze; consequently, to distinguish what is similar, what dissimilar; to discover proportion and disproportion, is the office of the understanding.

If the physiognomist has not an accurate, superior,

and extended understanding, he will neither be able rightly to observe, nor to compare and class his observations, much less to draw the necessary conclusions. Physiognomy is the highest exercise of the understanding the logic of corporeal varieties.

To the clearest and profoundest understanding, the true physiognomist unites the most lively, strong, comprehensive imagination, and a fine and rapid wit. Imagination is necessary to impress the traits with exactness, so that they may be renewed at pleasure; and to range the pictures in the mind as perfectly as if they still were visible, and with all possible order.

A keen penetration is indispensable to the physiognomist, that he may easily perceive the resemblance that exists between objects. Thus, for example, he sees a head or forehead possessed of certain characteristic marks: these marks present themselves to his imagination, and a keen penetration discovers to what they are similar. Hence greater precision, certainty, and expression, are imparted to his images. He must have the capacity of uniting the approximation of each trait that he remarks, and be able to define the degree of this approximation. No one, who is not inexhaustibly copious in language, can become a physiognomist; and the highest possible copiousness is poor, comparatively with the wants of physiognomy. All that language can express the physiognomist must be able to express. He must be the creator of a new language, which must be equally precise and alluring, natural and intelligible.

Every production of art, taste, and mind; all vocabularies of all nations; all the kingdoms of nature, must obey his command, must supply his necessities.

The art of drawing is indispensable, if he would be

precise in his definitions and accurate in his decisious. Drawing is the first, most natural, and unequivocal language of physiognomy; the best aid of the imagination, the only means of preserving and communicating numberless peculiarities, shades, and expressions, which are not by words or any other mode to be described. The physiognomist who cannot draw hastily, accurately, and characteristically, will be unable to make, much less to retain or communicate, innumerable observations.

The knowledge of anatomy is indispensable to him; as also is physiology, or the science of the human body in health; not only that he may be able to remark any disproportion, as well in the solids as in the muscular parts, but that he may likewise be capable of naming these parts in his physiognomical language. He must also be acquainted with the temperament of the human body; not only its different colours and appearances, occasioned by the mixture of the blood, but also the constituent parts of the blood itself, and their different proportions. Still more especially must be understood the external symptoms of the constitution, relative to the nervous system; for on this depends more than even on the knowledge of the blood.

What an extensive knowledge ought he to have of the human heart and the manners of the world! How thoroughly ought he to inspect, to feel himself! That most essential, yet most difficult of all knowledge, to the physiognomist, ought to be possessed by him in all possible perfection. In proportion only as he knows himself will he be enabled to know others.

Not only is this self-knowledge, this studying of man, by the study of his own heart, with the genealogy and consanguinity of inclinations and passions, their various symptoms and changes, necessary to the physiognomist for the foregoing causes, but also for an additional reason.

"The peculiar shades," I here cite the words of one of the critics on my first essay, "the peculiar shades of feeling, which most affect the observer of any object, frequently have relation to his own mind, and will be soonest remarked by him in proportion as they sympathize with his own powers. They will affect him most, according to the manner in which he is accustomed to survey the physical and moral world. Many, therefore, of his observations are applicable only to the observer himself; and, however strongly they may be conceived by him, he cannot easily impart them to others. Yet these minute observations influence his judgment. For this reason the physiognomist must, if he knows himself, which he in justice ought to do before he attempts to know others, once more compare his remarks with his own peculiar mode of thinking, and separate those which are general from those which are individual, and appertain to himself." I shall make no commentary on this important precept. I shall here only repeat, that an accurate and profound knowledge of his own heart is one of the most essential qualities in the character of the physiognomist.

Reader, if thou hast not often blushed at thyself, even though thou shouldest be the best of men, for the best of men is but man; if thou hast not often stood with downcast eyes in presence of thyself and others; if thou hast not dared to confess to thyself, and to confide to thy friend, that thou art conscious the seeds of every vice are latent in thy heart; if, in the gloomy calm of solitude, having no witness but God and thy own conscience, thou hast not a thousand times sighed and sorrowed for thyself; if thou wantest the power to observe the progress of the passions from their very commencement; to examine what the impulse was which determined thee to good or ill, and to avow the motive to God and thy friend, to whom thou mayest thus confess thyself, and who also may disclose the recesses of his soul to thee; a friend who shall stand before thee the representative of man and God, and in whose estimation thou also shalt be invested with the same sacred character; a friend in whom thou mayest see thy very soul, and who shall reciprocally behold himself in thee: if, in a word, thou art not a man of worth, thou never canst learn to observe or know men well; thou never canst be, never wilt be, worthy of being a good physiognomist. If thou wishest not that the talent of observation should be a torment to thyself, and an evil to thy brother, how good, how pure, how affectionate, how expanded ought thy heart to be! How mayest thou ever discover the marks of benevolence and mild forgiveness, if thou thyself art destitute of such gifts? How, if philanthropy does not make thine eye active, how mayest thou discern the impressions of virtue, and the marks of the sublimest sensations? How often wilt thou overlook them in a countenance disfigured by accident! Surrounded thyself by mean passions, how often will such false observers bring false intelligence! Put far from thee self-interest, pride, and envy, otherwise "thine eye will be evil, and thy whole body full of darkness." Thou wilt read vices on the forehead whereon virtue is written, and wilt accuse others of those errors and failings of which thy own heart accuses thee. Whoever bears any resemblance to thine enemy, will by thee be accused of all those failings and vices with which thy enemy is loaded by thy own partiality and self-love. Thine eye will overlook the beauteous traits and magnify the discordant. Thou wilt behold nothing but caricature and disproportion.

But, to draw to a conclusion: the physiognomist should know the world; he should have intercourse with all manner of men, in all various ranks and conditions; he should have travelled, should possess extensive knowledge, a thorough acquaintance with artists, mankind, vice, and virtue, the wise and the foolish, and particularly with children; together with a love of literature, and a taste for painting and the other imitative arts. I say, can it need demonstration that all those and much more are to him indispensable? To sum up the whole: to a well-formed, well-organized body, the perfect physiognomist must unite an acute spirit of observation, a lively fancy, an excellent judgment, and, with numerous propensities to the arts and sciences, a strong, benevolent, enthusiastic, innocent heart; a heart confident in itself, and free from the passions inimical to man. No one, certainly, can read the traits of magnanimity, and the high qualities of the mind, who is not himself capable of magnanimity, honourable thoughts, and sublime actions.

Thus have I pronounced judgment against myself in writing these characteristics of the physiognomist. Not false modesty, but conscious feeling, impels me to say, that I am as distant from the true physiognomist as heaven is from earth. I am but the fragment of a physiognomist, as this work is but the fragment of a system of physiognomy.

CHAPTER XVII.

Lavater's own Remarks on National Physiognomy.

IT is undeniable that there is national physiognomy as well as national character. Whoever doubts of this can never have observed men of different nations, nor have compared the inhabitants of the extreme confines of any two. Compare a Negro and an Englishman, a native of Lapland and an Italian, a Frenchman and an inhabitant of Terra del Fuego. Examine their forms, countenances, characters, and minds. Their difference will be easily seen, though it will sometimes be very difficult to describe it scientifically.

It seems to me probable that we shall discover what is national in the countenance better from the sight of an individual at first than of a whole people; at least, so it appears to me from my own experience. Individual countenances discover more the characteristic of a whole nation, than a whole nation does that which is national in individuals. The following infinitely little is what I have hitherto observed from the foreigners with whom I have conversed, and whom I have noticed, concerning national character.

I am least able to characterise the French. They have no trait so bold as the English, nor so minute as the Germans. I know them chiefly by their teeth and their laugh. The Italians I discover by the nose, small eyes, and projecting chin. The English by their fore-heads and eyebrows. The Dutch by the rotundity of the head, and the weakness of the hair. The Germans by the angles and wrinkles round the eyes and in the cheeks. The Russians by the snub nose, and their light-coloured or black hair.

I shall now say a word concerning Englishmen in particular. Englishmen have the shortest and best arched foreheads; that is to say, they are arched only upwards, and, towards the eyebrows, either gently recline or are rectilinear. They very seldom have pointed, but often round, full, medullary noses; the Quakers and Moravians excepted, who, wherever they are found, are generally thin-lipped. Englishmen have large, well-defined, beautifully curved lips. They have also a round full chin; but they are peculiarly distinguished by the eyebrows and eyes, which are strong, open, liberal, and steadfast. The outline of their countenance is in general great, and they never have those numerous, infinitely minute traits, angles, and wrinkles, by which the Germans are so especially distinguished. Their complexion is fairer than that of the Germans.

All English women, whom I have known personally or by portrait, appear to be composed of marrow and nerve. They are inclined to be tall, slender, soft, and as distant from all that is harsh, rigorous, or stubborn, as heaven is from earth.

The Swiss have generally no common physiognomy, or national character, the aspect of fidelity excepted. They are as different from each other as nations the most remote. The French Swiss peasant is as distinct as possible from the peasant of Appenzel. It may be that the eye of a foreigner would better discover the general character of the nation, and in what it differs from the French or German than that of the native.

I find characteristic varieties in each canton of Switzerland. The inhabitants of Zurich, for instance, are middle-sized, more frequently meagre than corpulent, but usually one or the other. They seldom have ardent eyes, and the outline is not often grand or minute. The men are seldom handsome, though the youth are incomparably so; but they soon alter. The people of Bern are tall, straight, fair, pliable, and firm, and are most distinguished by their upper teeth, which are white, regular, and easily to be seen. The inhabitants of Basle, or Basil, are more rotund, full, and tense of countenance, the complexion tinged with yellow, and the lips open and flaccid. Those of Schafhausen are hard-boned. Their eyes are seldom sunken, but are generally prominent. The sides of the forehead diverge over the temples, the cheeks fleshy, and the mouth wide and open. They are commonly stronger built than the people of Zurich, though in the canton of Zurich there is scarcely a village in which the inhabitants do not differ from those of the neighbouring village, without attending to dress, which, notwithstanding, is also physiognomic.

I have seen many handsome, broad-shouldered, strong, burden-bearing men, round Wadenschweil and Oberreid. At Weiningen, two leagues from Zurich, I met a company of well-formed men, who were distinguished for their cleanliness, circumspection, and gravity of deportment.

An extremely interesting and instructing book might be written on the physiognomic character of the peasants of Switzerland. There are considerable districts where the countenances, the nose not excepted, are most of them broad, as if pressed flat with a board. This disagreeable form, wherever found, is consistent with the character of the people. What could be more instructive than a physiognomic and characteristic description of such villages, their mode of living, food, and occupation?

CHAPTER XVIII.

Extracts from Buffon on National Physiognomy.

TRAVERSING the surface of the earth, and beginning in the north, we find in Lapland, and on the northern coast of Tartary, a race of men small of stature, singular of form, and with countenances as savage as their manners.

These people have large flat faces, the nose broad, the pupil of the eye of a yellow-brown inclining to a black, the eyelids retiring towards the temples, the cheeks extremely high, the mouth very large, the lower part of the face narrow, the lips full and high, the voice shrill, the head large, the hair black and sleek, and the complexion brown or tanned. They are very small and squat, though meagre. Most of them are not above four feet, and hardly any exceed four feet and a half. The Borandians are still smaller than the Laplanders. The Samoiedes more squat, with large heads and noses, and darker complexions. Their legs are shorter, their knees more turned outwards, their hair is longer, and they have less beard. The complexion of the Greenlanders is darker still, and of a deep olive colour.

The women, among all these nations, are as ugly as the men; and not only do these people resemble each other in ugliness, size, and the colour of their eyes and hair, but they have similar inclinations and manners, and are all equally gross, superstitious, and stupid. Most of them are idolaters; they are more rude than savage, wanting courage, self-respect, and modesty.

If we examine the neighbouring people of the long slip of land which the Laplanders inhabit, we shall find they have no relation whatever with that race, excepting only the Ostiachs and Tongusians. The Samoiedes and the Borandians have no resemblance with the Russians; nor have the Laplanders with the Finlanders, the Goths, Danes, or Norwegians. The Greenlanders are alike different from the savages of Canada. The latter are tall and well made; and, though they differ very much from each other, yet they are still more infinitely different from the Laplanders. The Ostiachs seem to be Samoiedes, something less ugly and dwarfish, for they are small and ill-formed.

All the Tartars have the upper part of the countenance very large and wrinkled, even in youth; the nose short and gross, the eyes small and sunken, the cheeks very high, the lower part of the face narrow, the chin long and prominent, the upper jaw sunken, the teeth long and separated, the eyebrows large, covering the eyes, the eyelids thick, the face flat, their skin of an olive colour, and their hair black. They are of a middle stature, but very strong and robust; have little beard, which grows in small tufts like that of the Chinese, thick thighs, and short legs.

The Little or Nogais Tartars have lost a part of their ugliness by having intermingled with the Circassians. As we proceed eastward into free or independent Tartary, the features of the Tartars become something less hard, but the essential characteristics of their race ever remain. The Mogul Tartars, who conquered China, and who were the most polished of these nations, are at present the least ugly and ill-made; yet have they, like the others, small eyes, the face large and flat, little beard, but always black or red, and the nose short and compressed.

Among the Kergisi and Teheremisi Tartars there is a whole nation or tribe, among whom are very singularly beautiful men and women. The manners of the Chinese and Tartars are wholly opposite, more so than are their countenances and forms. The limbs of the Chinese are well proportioned, large, and fat. Their faces are round and capacious, their eyes small, their eyebrows large, their eyelids raised, and their noses little and compressed. They only have seven or eight tufts of black hair on each lip, and very little on the chin.

The natives of the coast of New Holland, which lies in sixteen degrees fifteen minutes of south latitude, and to the south of the isle of Timor, are perhaps the most miserable people on earth, and of all the human race most approach the brute animal. They are tall, upright, and slender. Their limbs are long and supple, their heads great, their forehead round, their eyebrows thick. and their eyelids half shut. This they acquire by habit in their infancy, to preserve their eyes from the gnats, by which they are greatly incommoded; and, as they never much open their eyes, they cannot see at a distance, at least and unless they raise the head as if they wished to look at something above them. They have large noses, thick lips, and wide mouths. It should seem that they draw the two upper fore teeth, for neither man nor woman, young nor old, have these teeth. They have no beard; their faces are long and very disagreeable, without a single pleasing feature; their hair not long and sleek, like that of most of the Indians, but short, black, and curly, like the hair of the Negroes. Their skin is black, and resembles that of the Indians of the coast of Guinea.

Let us now examine the natives inhabiting a more

temperate climate, and we shall find that the people of the northern provinces of the Mogul empire, Persia, the Armenians, Turks, Georgians, Mingrelians, Circassians, Greeks, and all the inhabitants of Europe, are the handsomest, wisest, and the best formed of any on earth; and that, though the distance between Cachemire and Spain, or Circassia and France, is very great, there is still a very singular resemblance between people so far from each other, but situated in nearly the same latitude. The people of Cachemire are renowned for beauty, are as well formed as the Europeans, and have nothing of the Tartar countenance, the flat nose, and the small pig's eyes, which are so universal among their neighbours.

The complexion of the Georgians is still more beautiful than that of Cachemire; no ugly face is found in the country, and nature has endowed most of the women with graces which are nowhere else to be discovered. The men also are very handsome, have natural understanding, and would be capable of arts and sciences, did not their bad education render them exceedingly ignorant and vicious; yet with all their vices the Georgians are civil, humane, grave, and moderate; they seldom are under the influence of anger, though they become irreconcilable enemies having once entertained hatred.

The Circassians and Mingrelians are equally beautiful and well formed. The lame and the crooked are seldom seen among the Turks. The Spaniards are meagre, and rather small; they are well shaped, have fine heads, regular features, good eyes, and well-arranged teeth; but their complexions are dark, and inclined to yellow. It has been remarked that in some provinces of Spain, as near the banks of the river Bidassoa, the people have exceedingly large ears.

M. Lavater here makes this digression: Can large ears hear better than small? I know one person with large rude ears, whose sense of hearing is acute, and who has a good understanding; but, him excepted, I have particularly remarked large ears to betoken folly; and that, on the contrary, ears inordinately small appertain to very weak, effeminate characters, or persons of too great sensibility.—Thus far Lavater, let us now return to Buffon.

Men with black or dark-brown hair begin to be rather uncommon in England, Flanders, Holland, and the northern provinces of Germany; and few such are to be found in Denmark, Sweden, and Poland. According to Linnæus the Goths are very tall, have sleek, light-coloured, silver hair, and blue eyes. The Finlanders are muscular and fleshy, with long and light yellow hair, the iris of the eye a deep yellow.

If we collect the accounts of travellers, it will appear that there are as many varieties among the race of negroes as the whites. They also have their Tartars and their Circassians. The blacks on the coast of Guinea are extremely ugly, and emit an insufferable scent. Those of Sofala and Mozambique are handsome, and have no ill smell. These two species of negroes resemble each other rather in colour than features. Their hair, skin, the odour of their bodies, their manners and propensities, are exceedingly different. Those of Cape Verd have by no means so disagreeable a smell as the natives of Angola. Their skin also is more smooth and black, their body better made, their features less hard, their tempers more mild, and their shape better.

The negroes of Senegal are the best formed, and best receive instruction. The Nagos are the most humane, the Mondongos the most cruel, the Mimes the most resolute, capricious, and subject to despair.

The Guinea negroes are extremely limited in their capacities. Many of them appear to be wholly stupid; or, never capable of counting more than three, remain in a thoughtless state if not acted upon, and have no memory; yet, bounded as is their understanding, they have much feeling, have good hearts, and the seeds of all virtue.

The Hottentots have all very flat and broad noses; but these they would not have, did not their mothers suppose it their duty to flatten the nose shortly after birth. They have also very thick lips, especially the upper; the teeth white, the eyebrows thick, the head heavy, the body meagre, and the limbs slender.

The inhabitants of Canada and all these confines, are rather tall, robust, strong, and tolerably well made, have black hair and eyes, very white teeth, tawny complexion, little beard, and no hair, or almost none, on any other part of the body. They are hardy and indefatigable in marching, swift of foot, alike support the extremes of hunger or excess in feeding; are daring, courageous, haughty, grave, and moderate. So strongly do they resemble the eastern Tartars in complexion, hair, eyes, the almost want of beard and hair, as well as in their inclinations and manners, that we should suppose them the descendants of that nation, did we not see the two people separated from each other by a vast ocean. They also are under the same latitude, which is an additional proof of the influence of climate on the colour, and even on the form of man.

CHAPTER XIX.

Some of the most remarkable Passages from an excellent Essay on National Physiognomy, by Professor Kant of Konigsberg.

THE supposition of Maupertuis, that a race of men might be established in any province, in whom understanding, probity, and strength should be hereditary, could only be realized by the possibility of separating the degenerate from the conformable births; a project which, in my opinion, might be practicable, but which, in the present order of things, is prevented by the wiser dispositions of nature, according to which the wicked and the good are intermingled, that, by the irregularities and vices of the former, the latent powers of the latter may be put in motion, and impelled to approach perfection. If nature, without transplantation or foreign mixture, be left undisturbed, she will, after many generations, produce a lasting race that shall ever remain distinct.

If we divide the human race into four principal classes, it is probable that the intermediate ones, however perpetuating and conspicuous, may be immediately reduced to one of these :—1. The race of Whites. 2. The Negroes. 3. The Huns (Monguls or Calmucs). 4. The Hindoos, or people of Hindostan.

External things may well be the accidental, but not the primary causes of what is inherited or assimilated. As little as chance, or physico-mechanical causes, can produce an organized body, as little can they add any thing to its power or propagation; that is to say, produce a thing which shall propagate itself by having a peculiar form or proportion of parts.

Man was undoubtedly intended to be the inhabitant of all climates and all soils. Hence the seeds of many internal propensities must be latent in him, which shall remain inactive or be put in motion according to his situation on the earth. So that, in progressive generations, he shall appear as if born for that particular soil in which he seems planted.

The air and the sun appear to be those causes which most influence the powers of propagation, and effect a durable development of germ and propensities; that is to say, the air and the sun may be the origin of a distinct race. The variations which food may produce must soon disappear on transplantation. That which affects the propagating powers must not act upon the support of life, but upon its original source, its first principle, animal conformation, and motion.

A man transplanted to the frigid zone must decrease in stature, since, if the power or momentum of the heart continues the same, the circulation must be performed in a shorter time, the pulse become more rapid, and the heat of the blood increased. Thus Crantz found the Greenlanders not only inferior in stature to the Europeans, but also that they had a remarkably greater heat of body. The very disproportion between the length of the body and the shortness of the legs, in the northern people, is suitable to their climate; since the extremes of the body, by their distance from the heart, are more subject to the attacks of cold.

The prominent parts of the countenance, which can less be guarded from cold, by the care of nature for their preservation, have a propensity to become more flat. The rising cheek-bone, the half-closed blinking eye, appear to be intended for the preservation of sight against the dry cold air, and the effusions of light from the snow, (to guard against which the Esquimaux use snow spectacles,) though they may be the natural effect of the climate, since they are found only in a smaller degree in milder latitudes. Thus gradually are produced the beardless chin, the flattened nose, thin lips, blinking eyes, flat countenances, red-brown complexion, black hair, and, in a word, the face of the Calmuc. Such properties, by continued propagation, at length form a distinct race, which continues to remain distinct even when transplanted into warmer climates.

The copper colour, or red-brown, appears to be as natural an effect of the aridity of the air, in cold climates, as the olive-brown of the alkaline and bilious juices in warm; without taking the native disposition of the American into the estimate, who appears to have lost half the powers of life, which may be regarded as the effect of cold.

The growth of the porous parts of the body must increase in the hot and moist climates. Hence the thick short nose and projecting lips. The skin must be oiled, not only to prevent excessive perspiration, but also imbibing the putrescent particles of the moist air. The surplus of the ferruginous or iron particles, which have lately been discovered to exist in the blood of man, and which, by the evaporation of the phosphoric acidities, of which all negroes smell so strong, being cast upon the retiform membrane, occasions the blackness which appears through the cuticle; and this strong retention of the ferruginous particles seems to be necessary in order to prevent the general relaxation of the parts. Moist warmth is peculiarly favourable to the growth of animals, and produces the negro, who, by the providence of nature, perfectly adapted to his climate, is strong, muscular, agile; but dirty, indolent, and trifling.

The trunk or stem of the root may degenerate; but this having once taken root, and stifled other germs, resists any future change of form, the character of the race having once gained a preponderance in the propagating powers.

CHAPTER XX.

Extracts from other Writers on National Physiognomy.— From Winkelmann's History of Art.—From the Recherches Philosophiques sur les Americains, by M. de Pauw.—Observations by Lintz.—From a Letter written by M. Fuessli.—From a Letter written by Professor Camper.

From Winkelmann's History of Art.

WITH respect to the form of man, our eyes convince us that the character of nation as well as of mind is visible in the countenance. As nature has separated large districts by mountains and seas, so likewise has she distinguished the inhabitants by peculiarity of features. In countries far remote from each other, the difference is likewise visible in other parts of the body, and in stature. Animals are not more varied, according to the properties of the countries they inhabit, than men are; and some have pretended to remark that animals even partake of the propensities of the men.

The formation of the countenance is as various as language—nay, indeed, as dialects—which are thus or thus various in consequence of the organs of speech. In cold countries the fibres of the tongue must be less flexible and rapid than in warm. The natives of Greenland, and certain tribes of America, are observed to want some letters of the alphabet, which must originate in the same cause. Hence it happens that the northern languages have more monosyllables, and are more clogged with consonants, the connecting and pronouncing of which is difficult, and sometimes impossible, to other nations.

A celebrated writer has endeavoured to account for the varieties of the Italian dialects, from the formation of the organs of speech. "For this reason," says he, "the people of Lombardy, inhabiting a cold country, have a more rough and concise pronunciation; the inhabitants of Florence and Rome speak in a more measured tone; and the Neapolitans, under a still warmer sky, pronounce the vowels more open, and speak with more fulness."

Persons well acquainted with various nations, can distinguish them as justly from the form of their countenance as from their speech. Therefore, since man has ever been the object of art and artists, the latter have constantly given the forms of face of their respective nations; and that art among the ancients gave the form and countenance of man, is proved by the same effect having taken place among the moderns. German, Dutch, or French, when the artists neither travel nor study foreign forms, can be known by their pictures as perfectly as Chinese or Tartars. After residing many years in Italy, Rubens continued to draw his figures as if he had never left his native land.

Another passage from Winkelmann.

The projecting mouths of the negroes, which they have in common with their monkeys, is an excess of growth, a swelling occasioned by the heat of the climate; like as our lips are swelled by heat or sharp saline moisture, and also in some men by violent passion. The small eyes of the distant northern and eastern nations, are in consequence of the imperfection of their growth. They are short and slender. Nature produces such forms the more she approaches extremes, where she has to encounter heat or cold. In the one she is prompter and exhausted, and in the other crude, never arriving at maturity. The flower withers in excessive heat, and, deprived of sun, is deprived of colour. All plants degenerate in dark and confined places.

Nature forms with greater regularity the more she approaches her centre, and in more moderate climates. Hence the Grecian and our own idea of beauty, being derived from more perfect symmetry, must be more accurate than the idea of those in whom, to use the expression of a modern poet, the image of the Creator is half defaced.

From the Recherches Philosophiques sur les Americains, by M. de Pauw.

The Americans are most remarkable, because that many of them have no eyebrows, and none have beards; yet we must not infer that they are enfeebled in the organs of generation, since the Tartars and Chinese have almost the same characteristics. They are far, however, from being very fruitful, or much addicted to love. True it is, the Chinese and Tartars are not absolutely beardless. When they are about thirty a small penciled kind of whisker grows on the upper lip, and some scattered hairs at the end of the chin.

Exclusive of the Esquimaux, who differ in gait, form, features, and manners, from other savages of North America, we may likewise call the Arkansans a variety, whom the French have generally named the handsome men. They are all tall and straight, have good features, without the least appearance of beards; have regular eyelids, blue eyes, and fine fair hair; while the neighbouring people are low of stature, have abject countenances, black eyes, the hair of the head black as ebony, and of the body thick and rough.

Though the Peruvians are not very tall, and generally thick set, yet they are tolerably well made. There are many, it is true, who by being diminutive are monstrous. Some are deaf, dumb, blind, and idiots; and others want a limb when born. In all probability, the excessive labour to which they have been subjected by the barbarity of the Spaniards, has produced such numbers of defective men. Tyranny has an influence on the very physical temperament of slaves. Their nose is aquiline, their forehead narrow, their hair black, strong, smooth, and plentiful; their complexion an olive-red, the apple of the eye black, and the white not very clear. They never have any beard, for we cannot bestow that name on some short straggling hairs which sprout in old age; nor have either men or women the downy hair which generally appears after the age of puberty. In this they are distinguished from all people on earth, even from the Tartars and Chinese. As in eunuchs, it is the character of their degeneracy.

Judging by the rage which the Americans have to

mutilate and disfigure themselves, we should suppose they were all discontented with the proportions of their limbs and bodies. Not a single nation has been discovered in this fourth quarter of the globe, which has not adopted the custom of artificially changing either the form of the lips, the hollow of the ear, or the shape of the head, by forcing it to assume an extraordinary and ridiculous figure.

There are savages whose heads are pyramidal or conical, with the top terminating in a point. Others have flat heads with large foreheads, and the back part flattened. This caprice seems to have been the most fashionable, at least it was the most common. Some Canadians had their heads perfectly spherical. Though the natural form of the head really approaches the circular, these savages who, by being thus distorted, acquired the appellation of bowl or bullet-head, do not appear less disgusting for having made the head too round, and perverted the original purpose of nature, to which nothing can be added, from which nothing can be taken away, without some essential error being the result, which is destructive to the animal.

In short, we have seen, on the banks of the Maragnon, Americans with square or cubical heads; that is to say, flattened on the face, the top, the temples, and the occiput, which appears to be the last stage of human extravagance.

It is not easy to conceive how it was possible to compress and mould the bones of the skull into so many various forms, without most essentially injuring the seat of sense and the organs of reason, or occasioning either madness or idiotism; since we so often have examples, that violent contusions in the region of the temples have occasioned lunacy, and deprived the sufferers of intellectual capacity. For it is not true, as ancient narratives have affirmed, that all Indians with flat or sugar-loaf heads were really idiots. Had this been the case there must have been whole nations in America either foolish or frantic, which is impossible even in supposition.

Observation by Lintz.

To me it appears very remarkable that the Jews should have taken with them the marks of their country and race to all parts of the world; I mean their short, black, curly hair, and brown complexion. Their quickness of speech, haste and abruptness in all their actions, appear to proceed from the same causes. I imagine the Jews have more gall than other men.

Extract from a Letter written by M. Fuessli, dated at Presburg.

My observations have been directed (says this great designer and physiognomist) not to the countenance of nations only; being convinced from numberless experiments that the general form of the human body, its attitude and manner, the sunken or raised position of the head between or above the shoulders, the firm, the tottering, the hasty, or slow walk, may frequently be less deceitful signs of this or that character, than the countenance separately considered. I believe it possible so accurately to characterize man, from the calmest state of rest to the highest gradation of rage, terror, and pain, that from the carriage of the body, the turn of the head, and gestures in general, we shall be able to distinguish the Hungarian, the

Sclavonian, the Illyrian, the Wallachian; and to obtain a full and clear conception of the actual, and in general the prominent, characteristics of this or that nation.

Extract of a Letter from Professor Camper.

It would be very difficult, if not impossible, to give you my particular rules for delineating various nations and ages with mathematical certainty, especially if I would add all that I have had occasion to remark concerning the beauty of the antiques. These rules I have obtained by constant observations on the skulls of different nations, of which I have a large collection, and by a long study of the antiques.

To draw any head accurately in profile takes me much time. I have dissected the skulls of people lately dead, that I might be able to define the lines of the countenance, and the angle of these lines with the horizon. I was thus led to the discovery of the maximum and minimum of this angle. I began with the monkey, proceeded to the Negro and the European, till I ascended to the countenances of antiquity, and examined a Medusa, an Apollo, or a Venus de Medicis. This concerns only the profile. There is another difference in the breadth of the cheeks, which I have found to be the largest among the Calmucs, and much smaller among the Asiatic Negroes. The Chinese, and inhabitants of the Molucca and other Asiatic islands, appear to me to have broad cheeks with projecting jawbones; the under jawbone in particular very high, and almost forming a right angle, which among Europeans is very obtuse, and still more so among the African Negroes.

I have not hitherto been able to procure a real skull of an American, and therefore cannot say any thing on that subject.

I am almost ashamed to confess that I have not yet been able accurately to draw the countenance of a Jew, although they are so very remarkable in their features; nor have I yet obtained precision in delineating the Italian face. It is generally true that the upper and under jaw of the European is less broad than the breadth of the skull, and that among the Asiatics they are much broader; but I have not been able to determine the specific differences between European nations.

By physiognomical sensations I have very frequently been able to distinguish the soldiers of different nations —the Scotchman, the Irishman, and the native of England; yet I have never been able to delineate the distinguishing traits. The people of our provinces are a mixture of all nations; but in the remote and separated cantons I find the countenance to be more flat, and extraordinarily high from the eyes upward.

CHAPTER XXI.

Extracts from the Manuscript of a Man of Literature at Darmstadt, on National Physiognomy.

ALL tribes of people who live in uncultivated countries, and consequently are pastoral, not assimilated in towns, would never be capable of an equal degree of cultivation with Europeans, though they did not live thus scattered. Were the shackles of slavery taken off, still their minds would eternally slumber; therefore whatever remarks we can make upon them must be pathognomonical (or physiognomical), and we must confine ourselves to their respective powers of mind, not being able to say much of their expression.

Such people as do not bear our badges of servitude, are not so miserable as we suspect. Their species of slavery is more supportable in their mode of existence. They are incomparably better fed than German peasants, and have neither to contend with the cares of providing, nor the excesses of labour. As their race of horses exceeds ours in strength and size, so do their people those among us who have, or suppose they have, property. Their wants are few, and their understanding sufficient to supply the wants they have. The Russian or Polish peasant is of necessity carpenter, tailor, shoemaker, mason, thatcher, &c.; and when we examine their performances we may easily judge of their capacities. Hence their aptitude at mechanical and handicraft professions, as soon as they are taught their principles. Invention of what is great they have no pretensions to; their mind, like a machine, is at rest when the necessity that sets it in motion no longer impels.

Of the numerous nations subject to the Russian sceptre, I shall omit those of the extensive Siberian districts, and confine myself to the Russians properly so called, whose countries are bounded by Finland, Eastland, Livonia, and the borders of Asia. These are distinguishable by prodigious strength, firm sinews, broad breast, and colossal neck, which in a whole ship's crew will be the same, resembling the Farnesian Hercules; by their black, broad, thick, rough, strong hair, head and beard; their sunken eyes, black as pitch;

their short forehead, compressed to the nose, with an arch. We often find thin lips, though in general they are pouting, wide, and thick. The women have high cheek bones, hollow temples, snub noses, and retreating arched foreheads, with very few traits of ideal beauty. Their power of propagation exceeds belief, and at a certain period of life both sexes become frequently corpulent.

The Ukranians, of whom most of the regiments of Cossacks are formed, dwell in the centre. They are distinguished among the Russians almost as the Jews are among Europeans. They generally have aquiline noses, and are nobly formed; amorous, yielding, crafty, and without strong passions; probably because, for some thousands of years, they have followed agriculture, have lived in society, had a form of government, and inhabit a fruitful country, in a moderate climate resembling that of France. Among all these people the greatest activity and strength of body are united. They are as different from the German boor as quicksilver is from lead; and how our ancestors could suppose them to be stupid is inconceivable.

Thus, too, the Turks resemble the Russians. They are a mixture of the noblest blood of Asia Minor with the more material and gross Tartar. The Natolian, of a spiritual nature, feeds on meditation: he will for days contemplate a single object, seat himself at the chessboard, or wrap himself up in the mantle of taciturnity. The eye, void of passion or great enterprise, abounds in all the penetration of benevolent cunning; the mouth eloquent; the hair of the head and beard, and the small neck, declare the flexibility of the man.

The Englishman is erect in his gait, and generally stands as if a stake were driven through his body. His nerves are strong, and he is the best runner. He is distinguished from all other men by the roundness and smoothness of his face. If he neither speak nor move, he seldom declares the capability and mind he possesses in so superior a degree. His silent eye seeks not to please. His hair, coat, and character, are alike smooth. Not cunning, but on his guard; and, perhaps, but little colouring is necessary to deceive him on any occasion. Like the bull-dog, he does not bark; but, if irritated, rages. As he wishes not for more esteem than he merits, so he detests the false pretensions of his neighbours, who would arrogate excellence they do not possess. Desirous of private happiness, he disregards public opinion, and obtains a character of singularity. His imagination, like a seacoal fire, is not the splendour that enlightens a region, but expands genial warmth. Perseverance in study, and pertinacity for centuries in fixed principles, have raised and maintained the British spirit, as well as the British government, trade, manufactures, and marine. He has punctuality and probity, not trifling away his time to establish false principles, or making a parade with a vicious hypothesis.

In the temperament of nations the French class is that of the sanguine. Frivolous, benevolent, and ostentatious, the Frenchman forgets not his inoffensive parade till old age has made him wise. At all times disposed to enjoy life, he is the best of companions. He pardons himself much; and therefore pardons others if they will but grant that they are foreigners, and he is a Frenchman. His gait is dancing, his speech without accent, and his ear incurable. His imagination pursues the consequences of small things with the rapidity of the second-hand of a stop-watch, but seldom gives those loud, strong, reverberating strokes which proclaim new discoveries to the world. Wit is his inheritance. His countenance is open, and at first sight speaks a thousand pleasant, amiable things. Silent he cannot be, either with eye, tongue, or feature. His eloquence is often deafening; but his good-humour casts a veil over all his failings. His form is equally distinct from that of other nations, and difficult to describe in words. No other man has so little of the firm or deep traits, or so much motion. He is all appearance, all gesture; therefore the first impression seldom deceives, but declares who and what he is. His imagination is incapable of high flights, and the sublime in all arts is to him offence. Hence his dislike to whatever is antique in art or literature, his deafness to true music, his blindness to the higher beauties of painting. His last most marking trait is, that he is astonished at every thing, and cannot comprehend how it is possible men should be other than they are at Paris.

The countenance of the Italian is soul, his speech exclamation, his motion gesticulation. His form is the noblest, and his country the true seat of beauty. His short forehead, his strong marked eyebones, the fine contour of his mouth, give a kindred claim to the antiquities of Greece. The ardour of his eyes denotes that the beneficent sun brings forth fruit more perfect in Italy than beyond the Alps. His imagination is ever in motion, ever sympathizing with surrounding objects, and as in the poem of Ariosto the whole works of creation are reflected, so are they generally in the national spirit. That power which could bring forth such a work, appears to me the general representative of genius. It sings all, and from it all things are sung. The sublime

in arts is the birthright of the Italian.   Modern religion
and politics may have degraded and falsified his character, may have rendered the vulgar faithless and crafty,
but the superior part of the nation abounds in the
noblest and best of men.

The Dutchman is tranquil, patient, confined, and
appears to will nothing.   His walk and eye are long
silent, and an hour of his company will scarcely produce
a thought.   He is little troubled by the tide of passions,
and he will contemplate unmoved the parading streamers
of all nations sailing before his eyes.   Quiet and com-
petence are his gods; therefore those arts alone which
can procure these blessings employ his faculties.   His
laws, political and commercial, have originated in that
spirit of security which maintains him in the possession
of what he has gained.   He is tolerant in all that relates
to opinion, if he be but left peaceably to enjoy his
property, and to assemble at the meeting-house of his
sect.   The character of the ant is so applicable to the
Dutch, that to this literature itself conforms in Holland.
All poetical powers, exerted in great works or small, are
foreign to this nation.   They endure pleasure from the
perusal of poetry, but produce none.   I speak of the
United Provinces, and not of the Flemings, whose jovial
character is in the midway between the Italian and
French.   A high forehead, half-open eyes, full nose,
hanging cheeks, wide open mouth, fleshy lips, broad
chin and large ears, I believe to be characteristic of
the Dutchman.

A German thinks it disgraceful not to know every
thing, and dreads nothing so much as to be thought a
fool.   Probity often makes him appear a blockhead.   Of
nothing is he so proud as of · honest moral under-

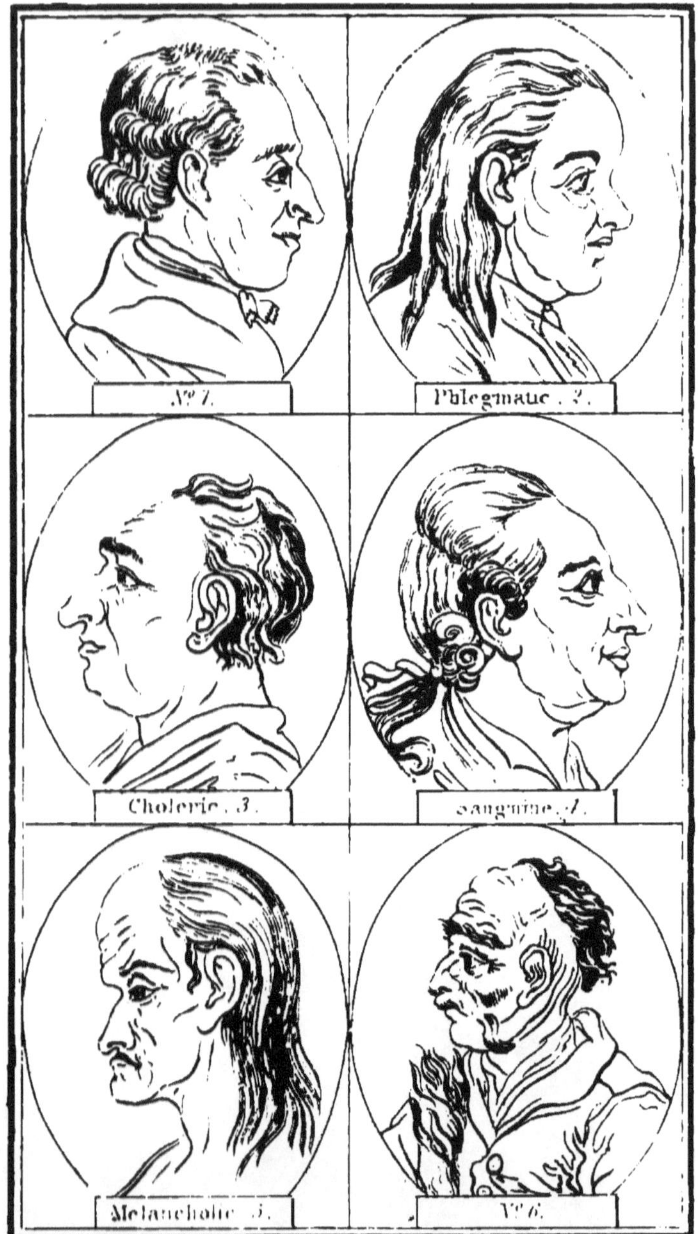

Nº 1.
Phlegmatic. 2.
Choleric. 3.
Sanguine. 4.
Melancholic. 5.
Nº 6.

standing. According to modern tactics he is certainly the best soldier, and the teacher of all Europe. He is allowed to be the greatest inventor, and often with so little ostentation, that foreigners have for centuries, unknown to him, robbed him of his glory. From the age of Tacitus, 'a willing dependent, he has exerted faculties for the service of his masters which others only exert for freedom and property. His countenance does not, like a painting in fresco, speak at a distance; but he must be sought and studied. His good nature and benevolence are often concealed under apparent moroseness, and a third person is always necessary to draw off the veil, and show him as he is. He is difficult to move, and without the aid of old wine is silent. He does not suspect his own worth, and wonders when it is discovered by others. Fidelity, industry, and secresy, are his principal characteristics. Not having wit, he indulges his sensibility. Moral good is the colouring which he requires in all acts. His epic and lyric spirit walk in unfrequented paths. Hence his great, and frequently gigantic sense, which seldom permits him the clear aspect of enthusisam, or the glow of splendour. Moderate in the use of this world's delights, he has little propensity to sensuality and extravagance; but he is therefore formal, and less social than his neighbour.

CHAPTER XXII.

Description of Plate IV.

Number 1.

WE may certainly call noses arched and pointed like this, witty; but the wit is restrained and moderated by

I

the acute understanding of the forehead, the sincere
religion of the eye, and the phlegm of the chin.

Number 2.

The descent from the nose to the lips in the phlegmatic countenance is unphlegmatic and heterogeneous;
nor does the curvature of the upper eyelids sufficiently
agree with the temperament. The outlines of the
phlegmatic are relaxed, obtuse, and hanging ; the outline
of the eye oblique. Be it understood there are other
tokens, and that all phlegmatic persons have not these
signs, although whoever has them is certainly phlegmatic. If the projecting under lip, which is itself a
sign of phlegm, since it is evidently a superabundance
and not a want of matter, be angular, and sharply
delineated, then it is a sign of choleric phlegm; that is
to say, of the ebullition of humidity. If it be flexible,
obtuse, powerless, and drooping, it is then pure phlegm.
The forehead, nose, chin, and hair, are here very phlegmatic.

Number 3.

The choleric ought to have a more angularly pointed
nose, and lips more sharply delineated. The character
of choler is much contained in the drawing of the eyes,
either when the pupil projects, and much of the under
part of the white is visible, or when the upper eyelid
retreats, so that it scarcely can be perceived; when the
eyes open, or when the eye is sunken, and the outlines
are very definite and firm, without much curvature. In
this example, the forehead, eyebrows, nose, chin, and
hair, are very choleric; but the upper part of the countenance more so than the under.

Number 4.

The sanguine needs but little correction, except that the nose ought to be a little farther from the mouth, and the eye not so choleric. The levity of the sanguine temperament waves, flutters upon the lip, which, however, at the bottom is too phlegmatic.

Number 5.

There ought to be a deeper cavity above the nose, and also of the jawbone, beside the ear, in this melancholic countenance. I have observed in many melancholic persons, that the nose declines towards the lips; nor have I seen this in any who were not sometimes inclined to the melancholic, who likewise have projecting under lips, and small, but not very round nor very fleshy chins.

There are melancholy persons with very sanguine temperaments; men of fine irritability and moral feelings, who are hurried into vices which they deeply abhor, and which they have not the power to withstand. The gloomy and dispirited character of such is perceptible in the eye that shuns examination, and the wrinkles of the forehead standing opposite to each other. Persons of a real melancholic temperament generally have their mouths shut, but the lips are always somewhat open in the middle. Many melancholy persons have small nostrils, and seldom well-arranged, clean, white teeth.

Number 6.

Strength and ardour, enterprise, courage, contempt of danger, fortitude of the irritated and irritable. This strength is rather oppressive than patient and enduring;

it proclaims its own qualities, respectable in a state of rest, terrible when roused.

CHAPTER XXIII.

Resemblance between Parents and Children.

THE resemblance between parents and children is very commonly remarkable. Family physiognomy is as undeniable as national. To doubt this is to doubt what is self-evident ; to wish to interpret it is to wish to explore the inexplicable secret of existence. Striking and frequent as the resemblance between parents and children is, yet have the relations between the characters and countenances of families never been inquired into. No one has, to my knowledge, made any regular observations on this subject. I must also confess that I myself have made but few with that circumstantial attention which is necessary. All I have to remark is as follows :—

When the father is considerably stupid, and the mother exceedingly the reverse, then will most of the children be endured with extraordinary understanding.

When the father is good, truly good, the children will in general be well-disposed ; at least most of them will be benevolent.

The son generally appears to inherit moral goodness from the good father, and intelligence from the intelligent mother ; the daughter partakes of the character of the mother.

If we wish to find the most certain marks of resemblance between parents and children, they should be observed within an hour or two after birth. We may then perceive whom the child most resembles in its formation. The most essential resemblance is usually afterwards lost, and does not perhaps appear again for many years ; or not till after death.

When children, as they increase in years, visibly increase in the resemblance of form and features to their parents, we cannot doubt but there is an increasing resemblance of character. How much soever the characters of children may appear unlike that of the parents they resemble, yet will this dissimilarity be found to originate in external circumstances ; and the variety of these must be great indeed, if the difference of character is not at length overpowered by the resemblance of form.

I believe that from the strongly delineated father the firmness and the kind (I do not say the form, but the kind) of bones and muscles are derived ; and from the strongly delineated mother the kind of nerves and form of countenance, if the imagination and love of the mother have not fixed themselves too deeply in the countenance of the man.

Certain forms of countenance, in children, appear for a time undecided whether they shall take the resemblance of the father or the mother ; in which case I will grant that external circumstances, preponderating love for the father or mother, or a greater degree of intercourse with either, may influence the form.

We sometimes see children who long retain a remarkable resemblance to the father, but at length change, and become more like the mother. I undertake not to expound the least of the difficulties that occur on this subject ; but the most modest philosophy may be permitted to compare uncommon cases with those which are known, even though they were inexplicable ; and this, I believe, is all that philosophy can and ought to do.

We know that all longings or mother's marks, and whatever may be considered as of the same nature, do not proceed from the father, but from the imagination of the mother. We also know that children most resemble the father only when the mother has a very lively imagination, and love for or fear of the husband. Therefore, as has before been observed, it appears that the matter and *quantum* of the power and of the life proceed from the father; and from the imagination of the mother, sensibility, the kind of nerves, the form, and the appearance.

There are certain forms and features of countenance which are long propagated, and others which as suddenly disappear. The beautiful and the deformed (I do not say forms of countenances, but what is generally supposed to be beauty and deformity) are not the most easily propagated; neither are the middling and insignificant; but the great and the minute are easily inherited, and of long duration.

Parents with small noses may have children with the largest and strongest defined; but the father or mother seldom, on the contrary, have a very strong, that is to say, large-boned nose, which is not communicated at least to one of their children, and which does not remain in the family, especially when it is in the female line. It may seem to have been lost for many years, but soon or late will again make its appearance, and its resemblance to the original will be particularly visible a day or two after death.

Where any extraordinary vivacity appears in the eyes of the mother, there is almost a certainty that these eyes will become hereditary; for the imagination of the mother is delighted with nothing so much as with the beauty of her own eyes. Physiognomical sensation has been hitherto much more generally directed to the eye than to the nose and form of the face; but if women should once be induced to examine the nose and form of the face, as assiduously as they have done their eyes, it is to be expected that the former will be no less strikingly hereditary than the latter.

Well-arched and short foreheads are easy of inheritance, but not of long duration; and here the proverb is applicable, *Quod citò fit, citò perit.* (Soon got, soon gone.)

It is equally certain and inexplicable, that some remarkable physiognomies of the most fruitful persons have been wholly lost to their posterity; and it is as certain and inexplicable that others are never lost. Nor is it less remarkable that certain strong countenances of the father or mother disappear in the children, and perfectly revive in the grandchildren.

As a proof of the powers of the imagination of the mother, we sometimes see that a woman shall have children by the second husband, which shall resemble the first at least in the general appearance. The Italians, however, are manifestly too extravagant when they suppose children who strongly resemble their father are base-born. They say that the mother, during the commission of a crime so shameful, wholly employs her imagination concerning the possibility of surprise by, and the image of, her husband. But were this fear so to act, the form of the children must not only have the very image of the husband, but also his appearance of rage and revenge, without which the adulterous wife could not imagine the being surprised by, or image of, her husband. It is this appearance, this rage that she fears, and not the man.

Natural children generally resemble one of their parents more than the legitimate.

The more there is of individual love, of pure, faithful, mild affection, the more is this love reciprocal and unconstrained between the father and mother—which reciprocal love and affection imply a certain degree of imagination, and the capacity of receiving impressions—the more will the countenances of the children appear to be composed of the features of the parents.

The sanguine, of all the temperaments, is the most easily inherited, and with it volatility; and, being once introduced, much industry and suffering will be necessary to exterminate this volatility.

The natural timidity of the mother may easily communicate the melancholy temperament of the father. Be it understood that this is easy if, in the decisive moment, the mother be suddenly seized by some predominant fear; and that it is less communicable when the fear is less hasty and more reflective. Thus we find those mothers who, during the whole time of their pregnancy, are most in dread of producing monstrous or marked children, because they remember to have seen objects that excited abhorrence, generally have the best formed and freest from marks; for the fear, though real, was the fear of reason, and not the sudden effect of an object exciting abhorrence rising instantaneously to sight.

When both parents have given a deep root to the choleric temperament in a family, it may probably be some centuries before it be again moderated. Phlegm is not so easily inherited, even though both father and mother should be phlegmatic; for there are certain moments of life when the phlegmatic acts with its whole power, though it acts thus but rarely, and these moments may and must have their effects; but nothing appears more easy of inheritance than activity and industry, when these have their origin in organization, and the necessity of producing alteration. It will be long before an industrious couple, to whom not only a livelihood, but business, is in itself necessary, shall not have a single descendant with the like qualities, as such mothers are generally prolific.

CHAPTER XXIV.

Remarks on the Opinions of Buffon, Haller, and Bonnet, concerning the Resemblance between Parents and Children.

THE theory or hypothesis of Buffon, concerning the cause of the human form, is well known, which Haller has abridged and more clearly explained in the following manner :—

"Both sexes have their semen, in which are active particles of a certain form. From the union of these the fruit of the womb arises. These particles contain the resemblance of all the parts of the father or mother. They are by nature separated from the rude and unformed particles of the human juices, and are impressed with the form of all the parts of the body of the father or mother. Hence arises the resemblance of children to their parents. This will account for the mixture of the features of father and mother in the children; for the spots of animals, when the male and female are of different colours; for the Mulatto produced by a Negro and a White; and for many other phenomena difficult to be resolved.

"Should it be asked how these particles can assume the internal structure of the body of the father, since they can properly be only the images of the hollow vessels, it may be answered that we know not all the powers of nature, and that she may have preserved to herself, though she has concealed it from her scholar, man, the art of making internally models and impressions which shall express the whole solidity of the model."

Haller, in his preface to Buffon's Natural History, has in my opinion indisputably confuted this system. But he has not only forborne to elucidate the resemblance between fathers and children, but while opposing Buffon, he has spoken so much on the natural physiological dissimilarity of the human body, that he appears to have denied this resemblance. Buffon's hypothesis offended all philosophy; and though we cannot entirely approve the theory of Bonnet, yet he has very effectually opposed the incongruities of Buffon, to which Buffon himself could scarcely give any serious faith. But he, as we shall soon see, has either avoided the question of resemblance between parents and children, or, in order to strengthen his own system, has rather sought to palliate than to answer difficulties.

BONNET, concerning organized Bodies.

"Are the germs of one and the same species of organized bodies perfectly like each other, or individually distinct? Are they only distinct in the organs which characterize sex, or have they a resembling difference to each other, such as we observe in individual substances of the same species of plants or animals?"

ANSWER.—"If we consider the infinite variety to be observed in all the products of nature, the latter will appear most probable. The differences which are to be observed in the individuals of the same species, probably depend more on the primitive form of the germs than in the connexion of the sexes."

On the resemblance between Children and their Parents.

"I must own that, by the foregoing hypothesis, I have not been successful in explaining the resemblance of features found between parents and children. But are not these features very ambiguous? Do we not suppose that to be the cause which probably is not so? The father is deformed, the son is deformed after the same manner, and it is therefore concluded that deformity is inherited. This may be true, but it may be false. The deformity of each may arise from very different causes, and these causes may be infinitely varied.

"It is not so difficult to explain hereditary diseases. We can easily conceive that defective juices may produce defective germs; and, when the same parts of the body are affected by disease in father or mother, and in child, this arises from the similar conformation of the parts, by which they are subject to like inconveniences. Besides, the misshapen body often originates in diseases being hereditary, which much diminishes the first difficulty. For, since the juices conducted to those parts are of a bad quality, the parts must be more or less ill formed, according as they are more or less capable of being affected by these juices."

REFLECTION.

Bonnet cannot find the origin of family likeness in his system. Let us, however, take this his system in the part where he finds the origin of hereditary disease. Shall the defective juices of father or mother very much alter the germ, and produce, in the very parts where the father or mother is injured, important changes of bad formation, more or less, according to the capability of the germ, and its power of resistance? And shall the healthy juices of the parent in no manner affect the germ? Why should not the healthy juices be as active as the unhealthy? Why should they not introduce the same qualities in miniature which the father and mother have in the gross; since the father and mother assimilate the nutriment they receive to their own nature, and since the seminal juices are the spiritual extract of all their juices and powers, as we have just reason to conclude from the most continued and accurate observations? Why should they not as naturally, and as powerfully, act upon the germ, to produce all possible resemblance? But which resemblance is infinitely varied, by differently changeable and changed circumstances; so that the germ continually preserves sufficient of its own original nature and properties, yet is always very distinct from the parents, and sometimes even seems to have derived very little from them, which may happen from a thousand accidental causes or changes.

Hence family resemblance and dissimilarity being summarily considered, we shall find that nature, wholly employed to propagate, appears to be entirely directed to produce an equilibrium between the individual power of the germ in its first formation, and the resembling power of the parents; but the originality of the first form of the germ may not wholly disappear before the too great power of resemblance to the parents, but that they may mutually concur, and both be subject to numberless circumstances, which may increase or diminish their respective powers, in order that the riches of variety, and the utility of the creature, and its dependence on the whole and the general Creator, may be the greater and more predominant.

Every observation on the resemblance between parents and children, which I have been enabled to make, convinces me that neither the theories of Bonnet nor Buffon give any systematic explanation of phenomena, the existence of which cannot be denied by the sophistry of hypothesis. Diminish the difficulties as much as we will, facts will still stare us in the face. If the germ exist preformed in the mother, can this germ, at that time, have physiognomy? Can it, at that time, resemble. the future, promiscuous, first, or second father? Is it not perfectly indifferent to either? or, if the physiognomical germ exist in the father, how can it sometimes resemble the mother, sometimes the father, often both, and often neither?

I am of opinion that something germ-like, or a whole capable of receiving the human form, must previously exist in the mother; but which is nothing more than the foundation of the future fatherly or motherly I know not what, and is the efficient cause of the future living fruit. This germ-like something, which, most especially constituted agreeable to the human form, is analogous to the nature and temperature of the mother, receives a peculiar individual personal physiognomy, according to the propensities of the father or mother, the disposition of the moment of conception, and probably of many other future decisive moments.

Still much remains to the freedom and predisposition of man. He may deprave or improve his state of the juices, he may calm or agitate his mind, may awaken every sensation of love, and by various modes increase or relax them. Yet I think that neither the nature of the bones, nor the muscles and nerves, consequently the character, depends on the physiognomical preformation preceding generation; at least they are far from depending on these alone, though I allow the organizable, the primitive form, always has a peculiar individuality, which is only capable of receiving certain subtile influences, and which must reject others.

CHAPTER XXV.

Observations on the New-born, the Dying, and the Dead.

I HAVE had opportunities of remarking in some children, about an hour after a birth attended with no difficulties, a striking though infantine resemblance in the profile to the profile of the father; and that in a few days this resemblance had nearly disappeared. The impression of the open air, nutriment, and perhaps of position, had so far altered the outlines that the child seemed entirely different.

Of these children I saw two dead, the one about six weeks, and the other about four years old; and, nearly twelve hours after death, I observed the same profile which I had before remarked an hour after birth; with this difference, that the profile of the dead child, as is natural, was something more tense and fixed than the living. A part of this resemblance, however, on the third day was remarkably gone.

One man of fifty, and another of seventy years of age, who fell under my observation while they were living and after death, appeared while living not to have the least resemblance to their sons, and whose countenances seemed to be of a quite different class; yet, the second day after death, the profile of the one had a striking resemblance to that of his eldest, and of the other to the profile of his third son, as much so as the profile of the dead children before mentioned resembled the living profile an hour after birth, stronger indeed, and, as a painter would say, harder. On the third day here also a part of the resemblance vanished.

I have uniformly observed, among the many dead persons I have seen, that sixteen, eighteen, or twenty-four hours after death, according to the disease, they have had a more beautiful form, better defined, more proportionate, harmonized, homogeneous, more noble, more exalted, than they ever had during life.

It occurred to me that there might be in all men an original physiognomy, subject to be disturbed by the ebb and flow of accident and passion; and is not this restored by the calm of death, like as troubled waters, being again left at rest, become clear?

I have observed some among the dying who had been the reverse of noble or great during life, and who some hours before their death, or perhaps some moments, (one was in a delirium,) have had an inexpressible ennobling of the countenance. Every body saw a new man; colouring, drawing, and grace—all was new, all bright as the morning; beyond expression noble and exalted; the most inattentive must see, the most insensible feel, the image of God. I saw it break forth and shine through the ruins of corruption, was obliged to turn aside and adore in silence. Yes, glorious God; still art thou there, in the weakest, most fallible men!

CHAPTER XXVI.

Of the Influence of Countenance on Countenance.

As the gestures of our friends and intimates become our own, so in like manner does their appearance. Whatever we love we would assimilate to ourselves; and whatever in the circle of affection does not change us into itself, that we change, as far as may be, into ourselves.

All things act upon us, and we act upon all things, but nothing has so much influence as what we love; and, among all objects of affection, nothing acts so forcibly as the countenance of man. Its conformity to our countenance makes it most worthy our affection. How might it act upon, how attract our attention, had it not some marks, discoverable or undiscoverable, similar to, at least of the same kind, with the form and feature of our own countenance!

Without, however, wishing farther to penetrate into what is impenetrable, or to define what is inscrutable, the fact is indubitable, that countenances attract countenances, and also that countenances repel countenances; that similarity of features between two sympathetic and affectionate men, increase with the development and mutual communication of their peculiar individual sensations. The reflection, if I may so say, of the person beloved remains upon the countenance of the affectionate.

The resemblance frequently exists only in a single point—in the character of mind and countenance. A resemblance in the system of the bones presupposes a resemblance of the nerves and muscles.

Dissimilar education may affect the latter so much, that the point of attraction may be invisible to the unphysiognomical eye. Suffer the two resembling forms to approach, and they will reciprocally attract and repel each other; remove every intervening obstacle, and nature will soon prevail. They will recognize each other; and rejoice in the flesh of their flesh, and the bone of their bone: with hasty steps will proceed to assimilate. Such countenances also, which are very different from each other, may communicate, attract, and acquire resemblance; nay, their likeness may become more striking than that of the former, if they happen to be more flexible, more capable, and to have greater sensibility.

This resemblance of features, in consequence of mutual affection, is ever the result of internal nature and organization, and, therefore, of the character of the persons. It ever has its foundation in a preceding, perhaps imperceptible resemblance, which might never have been animated or suspected, had it not been set in motion by the presence of the sympathetic being.

To give the character of those countenances which most easily receive and communicate resemblance, would be of infinite importance. It cannot but be known that there are countenances that attract all, others that repel all, and a third kind which are indifferent. The all-repelling render the ignoble countenances, over which they have continued influence, more ignoble. The indifferent allows no change. The all-attracting either receive, give, or reciprocally give and receive. The first change a little, the second more, the third most. "These are the souls (says Hemsterhuys the younger) which, happily or unhappily, add the most exquisite discernment to that excessive internal elasticity which occasions them to wish and feel immoderately; that is to say, the souls which are so modified, or situated, that their attractive force meets the fewest obstacles in its progress."

To study the influence of countenance, this intercourse of mind would be of the utmost importance. I have found the progress of resemblance most remarkable when two persons, the one richly communicative, the other apt to receive, have lived a considerable time together without foreign intervention; when he who gave had given all, or he who received could receive no more, physiognomical resemblance had attained its grand point.

Youth, irritable, and easy to be won, let me here say a word to thee. Oh! pause, consider, throw not thyself too hastily into the arms of an untried friend. A gleam of sympathy and resemblance may easily deceive thee. If the man who is thy second self have not yet appeared, be not rash, thou shalt find him at the appointed hour. Being found, he will attract thee to himself, will give and receive whatever is communicable. The ardour of his eyes will nurture thine, and the gentleness of his voice temper thy too piercing tones. His love will shine in thy countenance, and his image will appear in thee. Thou wilt become what he is, and yet remain what thou art. Affection will make qualities in him visible to thee, which never could be seen by an uninterested eye. This capability of remarking, of feeling what there is of divine in him, is a power which will make thy countenance assume his resemblance. ·

CHAPTER XXVII.

On the Influence of the Imagination on the Countenance.

I MUST not leave this subject wholly in silence; but . must content myself with saying only a few words, on which volumes might be written. The little, the nothing I have to say upon it, can only act as an inducement to deeper meditations on a theme so profound.

Our own countenance is actuated by imagination, rendering it in some measure resembling the beloved or hated image which is living, present, and fleeting before us, and is within the circle of our immediate activity. If a man deeply in love, and supposing himself alone, were ruminating on his beloved mistress, to whom his imagination might lend some charms which, if present, he would be unable to discover; were such a person observed by a man of penetration, it is probable that traits of the mistress might be seen in the countenance of this meditating lover. So might, in the cruel features of revenge, the features óf the enemy be read, whom imagination represents as present. And thus is the countenance a picture of the characteristic features of all persons exceedingly loved or hated.

It is possible that an eye less penetrating than that of an angel, may read the image of the Creator in the countenance of a truly pious person. He who languishes after Christ, the more lively, the more distinctly, the more sublimely, he represents to himself the very presence and image of Christ, the greater resemblance will his own countenance take of this image. The image of imagination often acts more effectually than the real presence; and whoever has seen him of whom we speak, the great *HIM*, though it were but an instantaneous glimpse, oh! how incessantly will the imagination reproduce his image in the countenance!

Our imagination also acts upon other countenances. The imagination of the mother acts upon the child; and hence men long have attempted to influence the imagination for the production of beautiful children. In my opinion, however, it is not so much the beauty of surrounding forms as the interest taken concerning forms in certain moments: and here, again, it is not so much the imagination that acts as the spirit, that being only the organ of the spirit. Thus, it is true that it is *the spirit that quickeneth the flesh*, and the image of the flesh (merely considered as such) *profiteth nothing.*

A look of love from the sanctuary of the soul, has certainly greater forming power than hours of deliberate contemplation of the most beautiful images. This forming look, if so I may call it, can as little be premeditatedly given, as any other naturally beautiful form can be imparted by a studious contemplation in the looking-glass. All that creates and is profoundly active in the inner man, must be internal, and be communicated from above; as I believe it suffers itself not to be occasioned, at least not by forethought, circumspection, or wisdom in the agent, to produce such effects. Beautiful forms or abortions are neither of them the work of art or study, but of intervening causes, of the quick-guiding providence, the pre-determining God.

Endeavour to act upon affection instead of the senses. If thou canst but incite love, it will of itself seek and find the powers of creation; but this very love must itself be innate before it can be awakened. Perhaps, however, the moment of this awakening is not in our power; and therefore to those who would, by plan and method, effect that which is in itself so extraordinary, and imagine they have had I know not what wise and physiological circumspection when they first awaken love, I might exclaim in the words of the enraptured songster: "I charge you, O ye daughters of Jerusalem, by the roes and the hinds of the field, that ye stir not up nor awake my love till he please." Here behold the forming genius—"Behold he cometh, leaping upon the moun·tains, skipping upon the hills, like a young hart."

Unforeseen moments, rapid as the lightning, in my opinion form and deform. Creation of every kind is momentaneous; the development, nutriment, change, improving, injuring, is the work of time, art, industry, and education. Creative power suffers itself not to be studied; creation cannot be premeditated. Marks may be moulded, but living essence, within and without resembling itself, the image of God, must be created, born, "not of the will of the flesh, nor of the will of man, but of God."

CHAPTER XXVIII.

The Effects of the Imagination on the Human Form.

THAT by the strength of imagination there are marks communicated by mothers to children during pregnancy, is equally true and comprehensible; that there are images, animals, fruit, or other substances, on the body of the child ; marks of the hand on the very parts where the pregnant person has been suddenly touched ; aversion to things which have occasioned disgust in the mother ; and a continued scurvy communicated to the child by the unexpected sight of a putrid animal. So many marks on the bodies of children, arising not from imaginary but real accidents, must oblige us to own that there is truth in that which is inconceivable. Therefore the imagination of the mother acts upon the child.

Of the innumerable examples that might be produced, I shall cite the two following :—

A woman during the time of her pregnancy was engaged in a card party, and only wanted the ace of spades to win all that was staked. It so happened, in the change of cards, that the so-much-wished-for ace was given her. Her joy at this success had such an effect upon her imagination, that the child of which she was pregnant, when born, had the ace of spades depicted in the apple of the eye, and without injury to the organ of sight.

The following anecdote is certainly true, and still more astonishing :—

A lady of Reinthal had, during her pregnancy, a desire to see the execution of a man who was sentenced to have his right hand cut off before he was beheaded. She saw the hand severed from the body, and instantly turned away and went home, without waiting to see the death that was to follow. This lady bore a daughter, who was living at the time this fragment was written, and who had only one hand. The right hand came away with the after-birth.

Moral marks as well as physical are perhaps possible.

I have heard of a physician who never failed to steal something from all the chambers through which he passed, which he would afterwards forget; and in the evening his wife, who searched his pockets, would find keys, snuff-boxes, étuis-cases, scissors, thimbles, spectacles, buckles, spoons, and other trinkets, which she restored to the owners. I have been likewise told of a child who, at two years of age, was adopted when begging at the door of a noble family, received an excellent education, and became a most worthy man except that he could not forbear to steal. The mothers of these two extraordinary thieves must, during pregnancy, have had an extraordinary desire to pilfer. It will be self-evident that, however insufferable such men are in a state of society, they are rather unfortunate than wicked. Their actions may be as involuntary as mechanical, and, in the sight of God, probably as innocent as the customary motions of our fingers when we tear bits of paper, or do any other indifferent, thoughtless action.

The moral worth of an action must be estimated by its intention, as the political worth must by its consequences. As little injury as the ace of spades, if the story be true, did to the countenance of the child, as little probably did this thievish propensity to the heart. Such a person certainly had no roguish look, no avaricious, downcast, sly, pilfering aspect, like one who is both soul and body a thief. I have not yet seen any man of such an extraordinary character, and therefore cannot judge of his physiognomy by experience; yet we have reason previously to conclude, that men so uncommon must bear some marks in their countenance of such deviation of character.

Those extraordinary large or small persons, by us called giants and dwarfs, should perhaps be classed among these active and passive effects of the imagination. Though giants and dwarfs are not properly born such, yet it is possible, however incomprehensible, that nature may first, at a certain age, suddenly enlarge or contract herself.

We have a variety of examples that the imagination appears not only to act upon the present, but on absence, distance, and futurity. Perhaps apparitions of the dying and the dead may be attributed to this kind of effect. Be it granted that these facts, which are so numerous, are true, and including not only the apparitions of the dead, but of the living, who have appeared to distant friends; after collecting such anecdotes, and adding others on the subject of presage and prediction, many philosophical conjectures will thence arise, which may probably confirm my following proposition.

The imagination, incited by the desire and languishing of love, or inflamed by passion, may act in distant places and times. The sick or dying person, for example, sighs after an absent friend who knows not of his sickness, or thinks of him at the time. The pining of the imagination penetrates, as I may say, walls, and appears in the form of the dying person, or gives signs of his presence similar to those which his actual presence gives. Is there any real corporeal appearances? No. The sick or dying person is languishing in his bed, and has never been a moment absent; therefore there is no actual appearance of him whose form has appeared. What, then, has produced this appearance? What is it that has acted thus at a distance on another's senses or imagination?—Imagination; but the imagination through the focus of passion.—How?—It is inexplicable. But who can doubt such facts who does not mean to laugh at all historical facts?

Is there any improbability that there may be similar moments of mind when the imagination shall act alike inexplicably on the unborn child? That the inexplicable disgusts, I will grant; I feel it perfectly. But is it not the same in the foregoing examples, and in every example of the kind? Like as cripples first become so many years after birth, which daily experience proves, may not, after the same inconceivable manner, the seeds of what is gigantic or dwarfish be the effects of the imagination on the fruit, which does not make its appearance till years after the child is born?

Were it possible to persuade a woman to keep an accurate register of what happened, in all the powerful moments of imagination during her state of pregnancy, she then might probably be able to foretell the chief incidents, philosophical, moral, intellectual, and physiognomical, which would happen to her child. Imagination, actuated by desire, love, or hatred, may, with more than lightning swiftness, kill or enliven, enlarge, diminish, or impregnate, the organized fœtus with the germ of enlarging or diminishing wisdom or folly, death or life, which shall first be unfolded at a certain time, and under certain circumstances. This hitherto unexplored, but sometimes decisive and revealed, creative and changing power of the soul, may be in its essence identically the same with what is called faith-working miracles, which latter may be developed and increased by external causes, wherever it exists, but cannot be communicated where it is not. A closer examination of the foregoing conjectures, which I wish not to be held for any thing more than conjectures, may perhaps lead to the profoundest secrets of physiognomy.

CHAPTER XXIX.

Essay by a late Learned Man of Oldenburg, M. Sturtz, on Physiognomy, interspersed with short Remarks by the Author.

"LIKE Lavater, I am perfectly convinced of the truth of physiognomy, and of the all significance of each limb and feature. Certain it is that the mind may be read in the lineaments of the body, and its motion in its features and their shades.

"Cause and effect, connection and harmony, exist through all nature; therefore between the external and internal of man. Our form is influenced by our parents, by the earth on which we walk, the sun that warms us with his rays, the food that assimilates itself with our substance, the incidents that determine the fortune of our lives. These all modify, repair, and chisel forth the body, and the marks of the tool are apparent both in body and in mind. Each arching, each sinuosity of the externals, adapts itself to the individuality of the internal. It is adherent and pliable, like wet drapery. Were the nose but a little altered, Cæsar would not be the Cæsar with whom we are acquainted.

"The soul being in motion, it shines through the body as the moon through the ghosts of Ossian, each passion throughout the human race has ever the same language."

From * east and to west, envy nowhere looks with the

* Those passages which are not marked with inverted commas are the observations of M. Lavater on the different parts of M. Sturtz's Essay.

satisfied air of magnanimity, nor will discontent appear like patience. Wherever patience is, there is it expressed by the same signs, as likewise are anger, envy, and every other passion.

"Philoctetes certainly expresses not the sensation of pain like a scourged slave. The angels of Raphael must smile more nobly than the angels of Rembrandt; but joy and pain still have each their peculiar expression : they act according to peculiar laws upon peculiar muscles and nerves, however various may be the shades of their expression; and the oftener the passion is repeated or set in motion, the more it becomes a propensity, a favourite habit, the deeper will be the furrows it ploughs.

"But inclination, capacity, modes and gradations of capacity, talents, and an ability for business, lie much more concealed. A good observer will discover the wrathful, the voluptuous, the proud, the discontented, the malignant, the benevolent, and the compassionate with little difficulty; but the philosopher, the poet, the artist, and their various partitions of genius, he will be unable to determine with equal accuracy. And it will be still more difficult to assign the feature or trait in which the token of each quality is seated, whether understanding be in the eyebone, wit in the chin, and poetical genius in the mouth."

Yet I hope, I believe—nay, I know—that the present century will render this possible. The penetrating author of this essay would not only have found it possible, but would have performed it himself, had he only set apart a single day to compare and examine a well-arranged collection of characters, either in nature or well-painted portraits.

"Whenever we meet with a remarkable man, our attention is always excited, and we are more or less empirical physiognomists. We perceive in the aspect, the mien, the smile, the mechanism of the forehead, sometimes wit, at others penetration. We expect and presage, from the impulse of latent sensation, very determined qualities from the form of each new acquaintance; and when this faculty of judging is improved by an intercourse with the world, we often suceed to admiration in our judgment on strangers.

"Can we call this feeling, internal unacquired sensation, which is inexplicable; or is it comparison, indication, conclusion from a character we have examined to another which we have not, and occasioned by some external resemblance? Feeling is the ægis of enthusiasts and fools, and, though it may often be conformable to truth, is still neither demonstration nor confirmation of truth; but induction is judgment founded on experience, and this way only will I study physiognomy.

"With an air of friendship I meet many strangers, with cool politness I recede from others, though there is no expression of passion to attract or to disgust. On farther examination, I always found that I have seen in them some trait either of a worthy or a worthless person with whom I was before acquainted.

"A child, in my opinion, acts from like motives when he evades, or is pleased with, the caresses of strangers, except that he is actuated by more trifling signs; perhaps by the colour of the clothes, the tone of the voice, or often by some motion which he has observed in the parent, the nurse, or the acquaintance."

This cannot be denied to be often the case, and indeed much more often than is commonly supposed; yet I make no doubt of being able to prove that there are, in nature and art, a multitude of traits, especially of the extremes of passionate as well as dispassionate faculties, which of themselves, and without comparison with former experiments, are with certainty intelligible to the most unpractised observer. I believe it to be incorporated in the nature of man, in the organization of our eyes and ears, that he should be actuated or repulsed by certain countenances as well as by certain tones. Let a child who has seen but a few men, view but the open jaws of a lion or a tiger, and the smile of a benevolent person, and his nature will infallibly shrink from the one, and meet the smile of benevolence with a smile; not from reason and comparison, but from the original feelings of nature. For the same reason we listen with pleasure to a delightful melody, and shudder at discordant shrieks. As little as there is of comparison or consideration on such an occasion, so is there equally little on the first of an extremely pleasing, or an extremely disgusting countenance.

" Mere sensation, therefore, is not the cause, since I have good reason, when I meet a person who resembles Turenne, to expect sagacity, cool resolution, and ardent enterprise. If, in three men, I find one possessed of the eyes of Turenne and the same marks of prudence; another with his nose and high courage; the third with his mouth and activity; I then have ascertained the seat where each quality expresses itself, and am justified in expecting similar qualities wherever I meet similar features.

" Had we, for centuries past, examined the human form, arranged characteristic features, compared traits, and exemplified inflections, lines, and proportions, and

had we added explanations to each, then would our
Chinese alphabet of the race of man be complete, and
we need but open it to find the interpretation of any
countenance. Whenever I indulge the supposition that
such an elementary work is not absolutely impossible,
I expect more from it than even Lavater. I imagine
we may obtain a language so rich and so determinate,
that it shall be possible, from description only, to restore
the living figure; and that an accurate description of
the mind shall give the outline of the body, so that the
physiognomist, studying some future Plutarch, shall
regenerate great men, and ideal form shall, with facility,
take birth from the given definition."

This is excellent; and, be the author in jest or earnest, this is what I entirely, without dreaming and most
absolutely, expect from the following century; for which
purpose, with God's good pleasure, I will hereafter hazard
some essays.

"With these ideal forms shall the chambers of future
princes be hung, and he who comes to solicit employ-
ment shall retire without murmuring, when it is proved
to him that he is excluded by his nose."

Laugh or laugh not, friends or enemies of truth, this
will, this must happen.

"By degrees, I imagine to myself a new and another
world, where error and deceit shall be banished."

Banished they would be were physiognomy the
universal religion, were all men accurate observers, and
were not dissimulation obliged to recur to new arts, by
which physiognomy, at least for a time, may be rendered
erroneous.

"We have to inquire whether we should therefore be
happier?"

We should certainly be happier, though the present contest between virtue and vice, sincerity and dissimulation, which so contributes to the development of the grand faculties of man, renders, as I may say, human virtue divine, exalting it to heaven.

"Truth is ever found in the medium : we will not hope too little from physiognomy, nor will we expect too much. Here torrents of objections break in upon me, some of which I am unable to answer. Do so many men in reality resemble each other? Is not the resemblance general ; and, when particularly examined, does it not vanish, especially if the resembling persons be compared feature by feature? Does it not happen that one feature is in direct contradiction to another ; that a fearful nose is placed between eyes which betoken courage ? "

In the firm parts, or those capable of sharp outlines, accidents excepted, I have never yet found contradictory features, but often have between the firm and the flexible, or the ground-form of the flexible and their apparent situation. By ground-form I mean to say that which is preserved after death, unless distorted by violent disease.

" It is by no means proved that resemblance of form universally denotes resemblance of mind. In families where there is most resemblance, there are often the greatest varieties of mind. I have known twins not to be distinguished from each other, between whose minds there was not the least similarity."

If this be literally true, I will renounce physiognomy, and whoever shall convince me of it, I will give him my copy of these fragments, and an hundred physiognomical drawings. Nor will I be my own judge : I leave it to the worthy author of this remark to choose three arbiters. Let them examine the fact accurately, and, if they confirm it, I will own my error. Shades, however, of these twin brothers will first be necessary. In all the experiments I have made, I declare, upon my honour, I have never made any such remark.

"In what manner shall we be able to explain the innumerable exceptions which almost overwhelm rule? I will only produce some from my own observation. Dr. Johnson had the appearance of a porter; not the glance of the eye, not any trait of the mouth, speak the man of penetration or of science."

When a person of our author's penetration and judgment thus affirms, I must hesitate, and say—He has observed this, I have not. But how does it happen that, in more than ten years' observation, I have never met any such example? I have seen many men, especially in the beginning of my physiognomical studies, whom I supposed to be men of sense, and who were not so; but never, to the best of my knowledge, did I meet a wise man whom I supposed a fool. In the frontispiece is an engraving of Johnson. Can a countenance more tranquilly fine be imagined, one that more possesses the sensibility of understanding, planning, scrutinizing? In the eyebrows only, and their horizontal position, how great is the expression of profound, exquisite, penetrating understanding?

"The countenance of Hume was that of a common man."

So says common report. I have no answer but that I suspect the aspect, or flexible features, on which most observers found their physiognomical judgment, have, as I may say, effaced the physiognomy of the bones; as, for example, the outline and arching of the forehead, to which scarcely one in a hundred direct their attention.

"Churchill had the look of a drover; Goldsmith of a simpleton; and the cold eyes of Strange do not betray the artist."

The greatest artists have often the coldest eyes. The man of genius and the artist are two persons. Phlegm is the inheritance of the mere artist.

"Who would say that the apparent ardour of Wille speaks the man who passed his life in drawing parallel lines?"

Ardour and phlegm are not incompatible: the most ardent men are the coolest. Scarcely any observation has been so much verified as this: it appears contradictory, but it is not. Ardent, quickly determining, resolute, laborious, and boldly enterprising men, the moment of ardour excepted, have the coolest of minds. The style and countenance of Wille, if the profile portrait of him in my possession be a likeness, have this character in perfection.

"It appears to me that Boucher, the painter of the graces, has the aspect of an executioner."

Truly so. Such was the portrait I received. But then, my good M. Sturtz, let us understand what is meant by these painters of the graces. I find as little in his works as in his countenance. None of the paintings of Boucher were at all to my taste. I could not contemplate one of them with pleasure, and his countenance had the same effect. I can now comprehend, said I, on the first sight of his portrait, why I have never been pleased with the works of Boucher.

"I once happened to see a criminal condemned to the wheel, who with satanic wickedness had murdered his benefactor, and who yet had·the benevolent and open countenance of an angel of Guido. It is not impossible to discover the head of a Regulus among guilty criminals, or of a vestal in the house of correction."

I can confirm this from experience. Far be contradiction from me on this subject. But such vicious persons, however hateful with respect to the appearance and effect of their actions, or even to their internal motives, were not originally wicked. Where is the pure, the noble, finely-formed, easily-irritated man, with angelic sensibility, who has not his devilish moments, in which, were not opportunity happily wanting, he might, in one hour, be guilty of some two or three vices which would exhibit him, apparently at least, as the most detestable of men? Yet may he be a thousand times better and nobler than numerous men of subaltern minds, held to be good, who never were capable of committing acts so wicked, for the commission of which they so loudly condemn him, and, for the good of society, are bound to condemn.

"Lavater will answer, 'Show me these men, and I will comment upon them, as I have done upon Socrates. Some small, often unremarked trait, will probably explain what appears to you so enigmatical.' But will not something creep into the commentary which never was in the text?"

Though this may be, yet it ought not to be the case. I will also grant that a man with a good countenance may act like a rogue; but, in the first place, at such a moment his countenance will not appear good; and, in the next, he will infinitely oftener act like a man of worth.

"Have we any right, from a known character, to draw

conclusions concerning one unknown? or, is it easy to discover what that being is, who wanders in darkness, and dwells in the house of contradiction; who is one creature to-day, and to-morrow the reverse?"

How true, how important is this! How necessary a beacon to warn and terrify the physiognomist!

"What judgment could we form of Augustus, if we were only acquainted with his conduct to Cinna? or of Cicero, if we knew him only from his consulate? How gigantic rises Elizabeth among queens; yet how little, how mean was the superannuated coquette! James II., a bold general and a cowardly king! Monk, the revenger of monarchs, the slave of his wife! Algernon Sydney and Russell, patriots worthy of Rome, sold to France! Bacon, the father of wisdom, a bribed judge! Such discoveries make us shudder at the aspect of man, and shake off friends and intimates like coals of fire from the hand. When such cameleon minds can be one moment great; at another contemptible, and alter their form, what can that form say?"

Their form shows what they may, what they ought to be, and their aspect in the moment of action what they are. Their countenance shows their power, and their aspect the application of their power. The expression of their littleness may probably be like the spots of the sun, invisible to the naked eye.

"Does not that medium through which we are accustomed to look tinge our judgment? Smellfungus views all objects through a blackened glass; another through a prism. Many contemplate virtue through a diminishing, and vice through a magnifying, medium."

How excellently expressed!

"A book written by Swift on physiognomy would

certainly have been very different from that of Lavater. National physiognomy is still a large uncultivated field. The families of the fair classes of the race of Adam, from the Esquimaux to the Greeks in Europe, and in Germany alone what varieties are there which can escape no observer? Heads bearing the stamp of the form of government, which ever will influence education; republican haughtiness, proud of its laws; the pride of the slave, who feels pride because he has the power of inflicting the scourges he has received; Greeks, under Pericles and under Hassan Pacha; Romans in a state of freedom, governed by emperors and governed by popes; Englishmen under Henry the Eighth and Cromwell. How have I been struck by the portraits of Hampden, Pym, and Vane! All produce varieties of beauty, according to the different nations."

'It is impossible for me to express how much I think myself indebted to the author of this spirited and energetic essay. How worthy an act was it in him, whom I had unintentionally offended, concerning whom I had published a judgment far from sufficiently noble, to send me this essay, with liberty to make what use of it I pleased! In such a manner, in such a spirit, may informations, corrections, or doubts be ever conveyed to me! Shall I need to apologize for having inserted it? or rather, will not most of my readers say, Give us more such.

CHAPTER XXX.

Quotations from Huart, with Remarks thereon.

1.

"MANY, who are really wise, often appear not to be so; and others who appear to be wise, are the reverse. Some, again, neither are nor appear to be wise, while others have the possession and appearance of wisdom."

A touchstone for many countenances.

2.

"The son is often brought in debtor to the great understanding of the father."

3.

"Wisdom in infancy denotes folly in manhood."

4.

"No aid can make those bring forth who are not pregnant."

We must not expect fruit where seed has not been sown. How advantageous, how important, would physiognomy become, were it, by being acquainted with every sign of intellectual and moral pregnancy, enabled to render aid to all the pregnant!

5.

"The external form of the head is what it ought to be, when it resembles a hollow globe slightly compressed at the sides, with a small protuberance at the forehead and back of the head. A very flat forehead, or a sudden descent at the back of the head, are no good tokens of understanding."

The profile of such a head, notwithstanding the com-pressure, would be more circular than oval. The profile of a good head ought to form a circle only when combined with the nose; therefore, without the nose it approaches much more to the oval than the circular. "A very flat forehead (says our author) is no good sign of understanding." True, if the flatness resembles that of the ox; but I have seen perfectly flat foreheads—let me be rightly understood, I mean flat only between and above the eyebrows—in men of great wisdom. Much, indeed, depends upon the position and curves of the outlines of the forehead.

6.

"Man has more brain than any animal. Were the quantity of the brain in two of the largest oxen compared to the quantity found in the smallest man, it would prove to be less."

7.

"Large oranges have thick skins and little juice. Heads of much bone and flesh have little brain. Large bones, with abundance of flesh and fat, are impediments to the mind."

8.

"The heads of wise persons are very weak, and susceptible of the most minute impressions."

Often, not always. And how wise? Wise to plan, but not to execute. Active wisdom must have harder bones. One of the greatest of this earth's wonders is a man in whom the two qualities are united, who has sensibility even to painful excess, and colossal courage to resist the impetuous torrent, the whirlpool, by which he shall be assailed. Such characters possess sensibility from the tenderness of bodily feeling; and strength not not so much in the bones as in the nerves.

9.

"A thick belly," says Galen, "a thick understanding." With equal truth or falsehood, I may add, a thin belly a thin understanding. Remarks so general, which would prove so many able and wise men to be fools, I value but little. A thick belly certainly is no positive token of understanding, it is rather positive for sensuality, which is detrimental to the understanding; but abstractedly, and unconnected with other indubitable marks, I cannot receive this as a general proposition.

10.

"Aristotle holds the smallest heads to be the wisest."

But this, with all reverence for so great a man, I think was spoken without reflection. Let a small head be imagined on a great body, or a great head on a small body, each of which may be found in consequence of accidents that excite or retard growth; and it will be perceived that, without some more definite distinction, neither the large nor the small head is, in itself, wise or foolish. It is true that large heads with short triangular foreheads are foolish, as are those large heads which are fat, and incumbered with flesh; but small, particularly round heads, with the like incumbrance, are intolerably foolish, and generally possess that which renders their intolerable folly more intolerable, a pretension to wisdom.

11.

"It is a good sign when a small person has a head somewhat large, and a large person has the head somewhat small."

Provided this extend no farther than *somewhat*, it may be supportable; but it is certainly for the best when the head is in such proportion to the body, that it is not remarkable either for its largeness or smallness.

12.

"Memory and imagination resemble the understanding, as a monkey does a man."

13.

"Whether the flesh be hard or tender, it is of no consequence to the genius, if the brain do not partake of the same quality; for experience tells us that the latter is very often of a different temperament to the other parts of the body. But when both the brain and the flesh are tender, they betoken ill to the understanding, and equally ill to the imagination."

14.

"Phlegm and blood are the fluids which render the flesh tender; and those being moist, according to Galen, render men simple and stupid. The fluids, on the contrary, which harden the flesh, are choler and melancholy, (or bile,) and these generate wisdom and understanding. It is, therefore, a much worse sign to have tender flesh than rough; and tender signifies a bad memory, with weakness of understanding and imagination."

It occurs to me that there is an intelligent tenderness of flesh, which announces much more understanding than do the opposite qualities of rough and hard. I can no more class coriaceous flesh as the characteristic of understanding, than I can tenderness of flesh, without being more accurately defined, as the characteristic of folly. It will be proper to distinguish between tender

and porous or spongy, and between rough and firm
without hardness.

15.

"We must examine the hair, if we wish to discover
whether the quality of the brain corresponds with the
flesh. If the hair be black, strong, rough, and thick, it
betokens strength of imagination and understanding."

I am of a different opinion. Let not this be expressed
in such general terms. At this moment I recollect a
very weak man, by nature weak, with exactly such hair.
This roughness (*sprodigheit*) is a fatal word, which, taken
in what sense it will, never signifies any good.

"But if the hair be tender and weak, it denotes
nothing more than goodness of memory."

Once more too little; it denotes a fine organization,
which receives the impression of images at least as
strongly as the signs of images.

16.

"When the hair is of the first quality, and we would
farther distinguish whether it betokens goodness of
understanding or imagination, we must pay attention
to the laugh. Laughter betrays the quality of the imagination."

I may venture to add, of the understanding, of the
heart, of power, love, hatred, pride, humility, truth, and
falsehood. Would I had artists who would watch for
and design the outlines of laughter! The physiognomy
of laughter would be the best of elementary books for
the knowledge of man. If the laugh be good, so is the
person. It is said of Christ that he never laughed. I
believe it; but, had he never smiled, he would not have

been human.   The smile of Christ must have contained the precise outline of brotherly love.

17.

" Heraclitus says, A dry eye, a wise mind."

18.

" We shall discover few men of great understanding who write a fine hand."

It might have been said, with more accuracy, a school-master's hand.

CHAPTER XXXI.

Remarks on an Essay on Physiognomy by Professor Lichtenberg.

MUCH intelligence, much ornament, and a mild diffusive eloquence, are blended in this essay. It is the work of a learned, penetrating, and, in many respects, highly meritorious person, who appears to possess much knowledge of men, and a large portion of the prompt spirit of 'observation. This essay merits the utmost attention and investigation. It is so interesting, so comprehensive, affords so much opportunity of remark for the physiognomist, and of remarks which I have yet to make, that I cannot avoid citing the most important passages, and submitting them to an unprejudiced and accurate examination.

It is far from my intention or wish to compare myself with the excellent author, to make any pretensions to his fanciful and brilliant wit, and still less to his learning and penetration. It is perhaps my wish, though I

dare not hope, to meet and answer him with the same
elegance as his polished mind and fine taste seem to
demand. I am sensible of those wants which are
peculiar to myself, and which must remain mine even
when I have truth on my side. Yet, worthy sir, be
assured that I shall never be unjust, and that, even
where I cannot assent to your observations, I shall never
forget the esteem I owe your talents, learning, and
merits.

We will now, in supposition, sit down in friendship
with your essay before us, and with that benevolence
which is most becoming men, philosophers in particular, explain our mutual sentiments concerning nature
and truth.

ON PHYSIOGNOMY.

" Certainly," says our author, " the freedom of thought,
and the very recesses of the heart, were never more
severely scrutinized than in the present age."

I cannot help thinking that, at the very beginning,
an improper point of view is taken, which may probably
lead the author and reader astray through the whole
essay. For my own part, at least, I know of no attacks
on the freedom of thought, or the secret recesses of the
heart. It is universally known that my labours have
been less directed to this than to the knowledge of predominant character, capacities, talents, powers, inclinations, activity, genius, religion, sensibility, irritability,
and elasticity of men in general, and not to the discovery
of actual and present thought. As far as I am con
cerned, the soul may and can, in our witty author's own
words, " brood as secretly over its treasures as it might
have done centuries ago ; may as tranquilly smile at the

progress of all Babylonian works, at all proud assailants of heaven, convinced that, long before the completion of their work, there shall be a confusion of tongues, and the master and the labourers shall be scattered."

I should enjoy the laugh as much as any one, at the arrogance of that physiognomist who should pretend to read in the countenance the most secret thoughts and motions of the soul at any given moment, although there are moments in which they are legible to the most unpractised physiognomist.

I am also of opinion that the secrets of the heart belong to pathognomy, to which· I direct my attention much less than to physiognomy; of which the author says, more wittily than truly, "it is as unnecessary to write as on the art of love."

The author is very right in reminding us, "that we ought to seek physiognomical instruction from known characters with great caution, and even diffidence."

Our author then says, "Whether physiognomy, in its utmost perfection, would promote philanthropy, is at least questionable."

I confidently answer unquestionable, and I hope immediately to induce the reasonable and philanthropic author to say the same. Physiognomy, in its utmost perfection, must mean the knowledge of men in its utmost perfection. And shall not this promote the love of man? or, in other words, shall it not discover innumerable perfections which the half physiognomist, or the unphysiognomist, are unable to discover? Noble and penetrating friend of man, while writing this you had forgotten what you had so truly, so beautifully said, "that the most hateful deformity might, by the aid of virtue, acquire irresistible charms:" and to whom more irresistible, more legible, than to the perfect physiognomist? Irresistible charms certainly promote not hatred, but love. From my own experience, I can sincerely declare that the improvement of my physiognomical knowledge has extended and increased the power of love in my heart.

Though this knowledge may sometimes be the author of affliction, still it is ever true that the affliction occasioned by certain countenances, endears, sanctifies, and renders enchanting whatever is noble and lovely, which often glows in the human countenance like embers among ashes. My attention to the discovery of this secret goodness is increased, and the object of my labours is its increase and improvement; and how do esteem and love extend themselves wherever I perceive a preponderance of goodness! On a more accurate observation, the very countenances that afflict me, and which for some moments incense me against humanity, do but increase a tolerant and benevolent spirit; for I then discern the load and the nature of that sensuality against which they have to combat.

All truth, all knowledge of what is, of what acts upon us, and on which we act, promotes general and individual happiness. Whoever denies this is incapable of investigation. The more perfect this knowledge is, the greater are its advantages. Whatever profits, whatever promotes happiness, promotes philanthropy. Where are happy men to be found without philanthropy? Are such beings possible? Were happiness and philanthropy to be destroyed or lessened by any perfect science, truth would war with truth, and eternal wisdom with itself.

He who can seriously maintain that a perfect science

may be detrimental to human society, or may not promote philanthropy, (without which happiness among men cannot be supposed,) is certainly not a man in whose company our author would wish to philosophize, as he certainly will, with me, assume it as an axiom that "the nearer truth the nearer happiness." The more our knowledge and judgment resemble the knowledge and judgment of the Deity, the more will our philanthropy resemble the philanthropy of the Deity. He who knows how man is formed, who remembers that he is but dust, is the most tolerant friend of man.

I believe angels to be better physiognomists and more philanthropic than men, though they may perceive in us a thousand failings and imperfections which may escape the most penetrating eye of man. God, having the most knowledge of spirit, is the most tolerant of spirits. And who was more tolerant, more affectionate, more lenient, more merciful than thou, who *needest not that any should testify of man, for thou knewest what was in man ?*

"It is certain that the industrious, the insinuating and active blockheads in physiognomy, may do much injury to society."

Be assured, my worthy sir, it is my earnest desire, my known endeavour, to deter such blockheads from studying physiognomy. This evil can be prevented only by accurate observation. True it is that every science may become dangerous when studied by the superficial and the foolish, and the very reverse when studied by the accurate and the wise. According to your own principles, therefore, we must agree in this, that none but the superficial, the blockhead, the fanatical enemy of knowledge and learning in general can wish to prevent "all investigation of physiognomical principles;" none but such a person "can oppose physiognomical labours; none but a blockhead will suppose it unworthy and impracticable in these degenerate days to awaken sensibility and the spirit of observation, or to improve the arts and the knowledge of men." To grant all this as you, sir, do, and yet to speak with bitterness against physiognomy and physiognomists, I call sowing tares among the good seed.

Our author next proceeds to distinguish between physiognomy and pathognomy. "Physiognomy (he defines to be) a capability of discovering the qualities of the mind and heart from the form and qualities of the external parts of the body, especially the countenance, exclusive of all transitory signs of the motion of the mind; and pathognomy, the whole semeiotica of the passions, or the knowledge of the natural signs of the motions of the mind, according to all their gradations and combinations."

I entirely agree with this distinction, and likewise subscribe to these given definitions.

It is in the next place asked, Is there physiognomy? is there pathognomy? To the latter the author justly replies, "This no man ever yet denied; for what would all theatrical representations be without it? The language of all ages and nations abounds with pathognomical remarks, and with which they are inseparably interwoven."

However, after reading the work several times, I cannot discover whether the author does or does not grant the reality of physiognomy. In one passage the author very excellently says, "No one will deny that in a world where all thing are cause and effect, and where miracles are not to be found, each part is a mirror of the whole. We are often able to conclude from what is near to what is distant, from what is visible to what is invisible, from the present to the past and the future. Thus the history of the earth is written, in nature's characters, in the form of each tract of country, of its sands, hills, and rocks. Thus each shell of the seashore proclaims the once included mind, connected, like the mind of man, with this shell. Thus also might the internal of man be expressed by the external on the countenance, concerning which we particularly mean to speak. Signs and traces of thought, inclination, and capacity, must be perceptible. How visible are the tokens impressed upon the body by trade and climate! yet what are trade and climate compared to the ever-active soul, creative in every fibre, of whose absolute legibility from all and to all no one doubts?"

The writer of the above excellent passage is the last person from whom I should have expected the following :—"What! the physiognomist will exclaim, can the soul of Newton reside in the head of a Negro, or an angelic mind in a fiendlike form?"

As little could I have expected this passage :—

"Talents, and the endowments of the mind in general, are not expressed by any signs in the firm parts of the head."

I have never in my life met with any thing more contradictory to nature, and to each other, than the foregoing and the following paragraphs :—

"If a pea were thrown into the Mediterranean, an eye more piercing than ours, though infinitely less penetrating than the eye of Him who sees all things, might perceive the effects produced on the coast of China." These are our author's very words.

And shall the whole living powers of the soul, "creative in every fibre," have no determinate influence on the firm parts, those boundaries of its activity, which first were yielding, and, acted upon, impressed by every muscle; which resemble each other in no human body, which are so various as characters and talents, and are as certainly different as the most flexible parts of man? Shall the whole powers of the soul, I say, have no determinate influence on these, or not by these be defined?

In order to avoid the future imputation of indulging the shallow stream of youthful declamation, instead of producing facts and principles deduced from experience, let us oppose experience to declamation, and facts to subtleties. But first a word, that we may perfectly remove a degree of ambiguity which I should not have expected from the accuracy of a mathematician.

"Why not," says our author—"why not the soul of Newton in the head of a Negro? Why not an angel mind in a fiend-like form? Who, reptile, empowered thee to judge of the works of God?"

Let us represent things in their proper light. We do not speak here of what God can do, but of what is to be expected from the knowledge we have of his works. We ask what the Author of order actually does, and not whether the soul of Newton can exist in the body of a Negro, or an angelic soul in a fiendlike form. The physiognomical question is, Can an angel's soul act the same in a fiendlike body as in the angelic body? or, in other words, Could the mind of Newton have invented the theory of light, residing in the head of a Negro, thus and thus defined? Such is the question.

Will you, sir, who are the friend of truth—will you answer, It might? You who have previously said of the world, "All things in it are cause and effect, and miracles are not to be found ?"

"I should indeed be a reptile judging the works of God, did I maintain its impossibility by miracle; but the question at present is not concerning miracles; it is concerning natural cause and effect."

After having thus stated the argument, permit me, sir, to decide it by quoting your own words: "Judas scarcely could be that dirty, deformed mendicant painted by Holbein. No hypocrite who associates with the good, betrays with a kiss, and afterwards hangs himself, has the look of Holbein's Judas. My experience leads me to suppose Judas must have been distinguished by an insinuating countenance and an ever-ready smile."

How true! how excellent! Yet what if I were to exclaim, "Who empowered thee, reptile, to judge of the works of God ?" What if I were to retort the following just remark, "Tell me first, why a virtuous mind is so often doomed to exist in an infirm body? Might not also, were it God's good pleasure, a virtuous man have a countenance like the beggarly Jew of Holbein, or any other that can be imagined ?"

Can this, however, be called wise or manly reasoning? How wide is the difference between suffering and disgusting virtue! or, is it logical to deduce that, because virtue may suffer, virtue may be disgustful? Is not suffering essential to virtue! To ask why virtue must suffer, is equivalent to asking why God has decreed that virtue should exist. Is it alike incongruous to admit that virtue suffers, and that virtue looks like vice? Virtue void of conflict, of suffering, or of self-denial, is not virtue accurately considered; therefore it is folly to ask, why must the virtuous suffer? It is in the nature

of things ; but it is not in the nature of things, not in the
relation of cause and effect, that virtue should look like
vice, or wisdom like foolishness. How, good sir, could
you forget what you have so expressively said, " There
is no durable beauty without virtue; and the most hateful deformity may, by the aid of virtue, acquire the
most irresistible charms? The author is acquainted
with several women whose example might inspire the
most ugly with hope."

What may be the infirmities of the virtuous we do
not inquire, nor whether a man of genius may become a
fool; we ask whether virtue, while existing, can look
like present vice, or actual folly like actual wisdom?
You, sir, who are so profound an inquirer into the
nature of man, will certainly never grant (who, indeed,
will?) that the soul of the beloved disciple of Christ
could, without a miracle, reside in the dirty, deformed
mendicant, the beggarly Jew of Holbein, and act as freely
in that as in any other body. Will you, sir, continue to
rank yourself, in your philosophical researches, with
those who, having maintained such senseless propositions, rid themselves of all difficulties by asking. " Who
empowered thee, reptile, to judge of the works of God ?"

Let us proceed to examine a few more passages.

" Our senses acquaint us only with the superficies,
from which all deductions are made. This is not very
favourable to physiognomy, for which something more
definite is requisite, since this reading of the superficies
is the source of all our errors, and frequently of our
ignorance."

So it is with us in nature : we absolutely can read
nothing more than the superficies. In a world devoid
of miracles, the external ever must have a relation to

the internal; and, could we prove all reading of the superficies to be false, what should we effect but the destruction of all human knowledge? All our inquiries produce only new superficies; all our truth must be the truth of the superficies. It is not the reading of the superficies that is the source of all our error; for, if so, we should have no truth; but the not reading, or, which is the same in effect, the not rightly reading.

If "a pea thrown into the Mediterranean Sea would effect a change in the superficies which should extend to the coast of China," any error that we might commit in our conclusions concerning the action of this pea, would not be because we read only the superficies, but because we cannot read the superficies.

"That we can only read the superficies is not very favourable to physiognomy, for which something more definite is requisite." Something more definite we have endeavoured to give, and wish to hear the objections of acute inquirers. But let facts be opposed to facts. Does not our author, by the expression "since the internal is impressed upon the external," seem to grant the possibility of this impression? And if so, does not the superficies become the index of the internal? Does he not thereby grant the physiognomy of the firm parts?

He proceeds to ask, "If the internal be impressed upon the external, is the impression to be discovered by the eyes of men?" Dare I trust my eyes that I have read such a passage in the writings of a philosopher?

We certainly see what we see. Be the object there or be it not, the question ever must be, Do we or do we not see? That we do see, and that the author, whenever he pleases, sees also, his essay is a proof, as are his other works. Be this as it may, I know not what would become of all our philosophers and philosophy, were we, at every new discovery of things, or the relations of things, to ask, Was this thing placed there to be discovered? With what degree of ridicule would our witty author treat the man who should endeavour to render astronomy contemptible by asking, "Though the wisdom of God is manifest in the stars, were the stars placed there to be discovered?

"Must not signs and effects which we do not seek, conceal and render those erroneous of which we are in search?"

The signs we seek are manifest, and may be known: they are the terminations of causes, therefore effects, therefore physiognomical expessions. The philosopher is an observer, an observer of that which is sought or not sought. He sees and must see that which presents itself to his eyes; and that which presents itself is the symbol of something that does not present itself. What he sees can only mislead him when he does not see rightly. If the conclusion be true, "that signs and effects which we do not seek, must conceal and render erroneous those of which we are in search," then ought we to seek no signs and effects, and thus all sciences vanish.

I have reason to hope that a person of so much learning as is our author, would not sacrifice all human sciences for the sole purpose of heaping physiognomy on a pile. I grant the possibility and facility of error is there; and this should teach us circumspection, should teach us to see the thing that is, without the addition of any thing that is not. But to wish, by any pretence, to divert us from seeing and observing, and to render inquiry contemptible, whether with rude or refined wit, would be the most ridiculous of all fanaticism. Such ridicule, in the mouth of a professed enemy of false philosophers, would be as vapid as false. I am indeed persuaded that my antagonist is not serious, and in earnest.

"Were the growth of the body (says the author) in the most pure of atmospheres, and modified only by the emotions of the mind, undisturbed by any external power, the ruling passion and the prevailing talent, I allow, might produce, according to their different gradations, different forms of countenance, like as different salts crystallize in different forms, when obstructed by no impediment. But is the body influenced by the mind alone, or is it not rather exposed to all the impulses of various contradictory powers, the laws of which it is obliged to obey? Thus each mineral, in its purest state, has its peculiar form; but the anomalies which its combination with others occasions, and the accidents to which it is subjected, often cause the most experienced to err when they would distinguish it by its form."

How strange is this simile! Salts and minerals compared to an organized body, internally animate! A grain of salt, which the least particle of water will instantaneously melt, to the human skull, which has defied misfortune and millions of external impressions for centuries! Dost thou not blush, Philosophy? Not to confine ourselves to the organization or the skulls of men and other animals, do we find that even plants, which have not the internal resistance, the elasticity of man, and which are exposed to millions of counteracting impressions from light, air, and other bodies, ever change their form in consequence of such causes? Which of them is ever mistaken for another by the botanist? The most violent accidents scarcely could effect such a change, so long as they should preserve their organization.

" Thus is the body mutually acted upon by the mind and external causes, and manifests not only our inclinations and capacities, but also the effects of misfortune, climate, diseases, food, and thousands of inconveniences to which we are subjected, not always in consequence of our vice, but often by accidents, and sometimes by our virtues."

Nobody can or will attempt to deny this. But is the foregoing question hereby answered? We are to attend to that. Does not our essayist himself say, " The body is acted upon by the mind and external causes?" Therefore not by external causes alone. May it not equally be affected by the internal energy or inactivity of the mind? What are we contending for? Has it not (if indeed the author be in earnest) the appearance of sophistry to oppose external to internal effects, and yet own the body is acted upon by both? And will you, sir, acute and wise as you are, maintain that misfortune can change a wise, a round, and an arched, into a cylindrical forehead; one that is lengthened into one that is square; or the projecting into the short retreating chin? Who can seriously believe and affirm that Charles XII., Henry IV., and Charles V., men who were undoubtedly subject to misfortunes if ever men were, thereby acquired another form of countenance, (we speak of the firm parts, not of scars,) and which forms denoted a different character to what each possessed previous to such misfortunes? Who will maintain that the noses of Charles XII. or Henry IV., denoting power of mind previous to their reverse of fortune, the one at Pultowa, the other by the hand of Ravaillac, suffered any change, and were debased to the insignificant pointed nose of a girl? Nature acts from within upon the bones; accident and suffering act on the nerves, muscles, and skin. If any accident attack the bones, who is so blind as not to remark such physical violence? The signs of misfortune are either strong or feeble: when they are feeble, they are effaced by the superior strength and power of nature; when strong, they are too visible to deceive, and by their strength and visibility warn the physiognomist not to suppose them the features of nature. By the physiognomist I mean the unprejudiced observer, who alone is the real physiognomist, and has the right to decide; not the man of subtlety, who is wilfully blind to experience.

"Are the defects which I remark in an image of wax always the defects of the artist, or are they not the consequences of unskilful handling, the sun's heat, or the warmth of the room?"

Nothing, dear friend of truth, is more easy to observe, in an image of wax, than the original hand of the master, although it should, by improper handling, accidental pressure, or melting, be injured. This example, sir, militates against yourself. If the hand of the master be visible in an image of wax, where it is so easily defaced, how much more perceptible must accident be in an organized body, so individually permanent? Instead of an image of wax, the simile, in my opinion, would be improved were we to substitute a statue; and in this every connoisseur can distinguish what has been broken, chopped, or filed off, as well as what has been added by a later hand. And why should not this be known in man? Why should not the original form of man be more distinguishable, in despite of accident, than the beauty and workmanship of an excellent statue which has been defaced?

"Does the mind, like an elastic fluid, always assume the form of the body? And if a flat nose were the sign of envy, must a man, whose nose by accident should be flattened, consequently become envious?"

The inquirer will gain but little, be this question answered in the negative or affirmative. What is gained were we to answer, "Yes; the soul is an elastic fluid, which always takes the form of the body?" Would it thence follow that the flattened nose has lost so much of its elasticity as would be necessary to propel the nose? or where would be the advantage should we reply, "No; all such comparisons are insignificant except to elucidate certain cases; we must appeal only to facts?"

But what would be answered to a less subtle and more simple question, Is there no example of the mind being injured by the maiming of the body? Has not a fractured skull, by compressing the brain, injured the understanding? Does not castration render the male half female?—But to answer wit with reason, says a witty writer, is like endeavouring to hold an eel by the tail.

We wholly subscribe to the affirmation, that "it is absurd to suppose the most beautiful mind is to be found in the most beautiful body, and the most deformed mind in the most deformed body."

We have already explained ourselves so amply on this subject, that being supposed to hold a contrary opinion appears incomprehensible. We only say there is a proportion and beauty of body which is more capable of superior virtue, sensibility, and action, than the disproportionate. We say with the author, "Virtue beautifies, vice deforms." We must cordially grant that honesty may be found in the most ugly, and vice in men of the most beautiful forms.

We cannot, however, help differing from him concerning the following assertion : "Our languages are exceedingly barren of physiognomical terms. Were it a true science, the language of the vulgar would have been proverbially rich in its terms. The nose occurs in a *hundred* proverbs and phrases, but always pathognomically, denoting past action, but never physiognomically, betokening character or disposition."

Instead of a *hundred*, I am acquainted with only one such phrase, *nasen rumfe*, to turn up the nose. *Homo obesæ, obtusæ naris*, said the ancients ; and, had they not said it, what could thence have been adduced, since we can prove *à posteriori* that the nose is a physiognomical sign of character ?

I have not learning sufficient, nor have I the inclination to cite sufficient proofs of the contrary from Homer, Suetonius, Martial, and an hundred others. That which is is, whether perceived by the ancients or not. Such dust might blind a school-boy, but not the eyes of a sage, who sees for himself, and who knows that each age has its measure of discovery, and that there are those who fail not to exclaim against all discoveries which were made by the ancients.

"I should be glad to know, (says our author,) not what man may become, but what he is."

I must confess that I wish to know both. Many vicious men resemble valuable paintings, which have been destroyed by varnish. Would you pay no attention to such a painting ? Is it wholly unworthy of you, though a connoisseur should assure you the picture is damaged, but there is a possibility of clearing away the varnish, as this master's colours are so strongly laid on, and so essentially good, that no varnish can penetrate deep enough, if we are but careful in bringing it away not to injure the picture? Is this of no importance? You observe the smallest change of position in the polar star. Days are dedicated to examine how many ages shall elapse before it will arrive at the nearest point of approach. I do not despise your labours. But is it of no importance to you, to fathers, mothers, guardians, teachers, friends, and statesmen, to inquire what a man may become, or what must be expected from this or that youth, thus and thus formed and educated? Many foolish people are like excellent watches, which would go well were the regulator but rectified.

Is the goodness of the mechanism of no consequence to you, although a skilful watchmaker should tell you, this was and is an excellent piece of workmanship, infinitely better than that which you see set with brilliants, which, I grant, will go well for a quarter of a year, but will then stop? Clean this, repair it, and straighten the teeth of this small wheel. Is this advice of no importance? Will you not be informed what it might have been, what it may yet probably be? Will you not hear of a treasure that lies buried, and, while buried, I own useless; but will you content yourself with the trifling interest arising from this or that small sum?

Is your attention paid only to the fruit of the present year, and which is perhaps forced? And do you neglect the goodness of a tree which, with attention, may bring forth a thousandfold, though under certain circumstances it may have brought forth none ? Have the hot blasts of the south parched up its black leaves, or has the storm blown down its half-ripened fruit, and will you therefore not inquire whether the root does not still flourish ?

I find I grow weary, and perhaps weary others, especially as I am more and more convinced that our pleasant author, at least hitherto, meant only to amuse himself. I shall therefore only produce two more contradictions which ought not to have escaped the author, and scarcely can escape any thinking reader.

He very properly says in one place, "pathognomical signs, often repeated, are not always entirely effaced, but leave physiognomical impressions. Hence originate the lines of folly, ever gaping, ever admiring, nothing understanding; hence the traits of hypocrisy; hence the hollowed cheek, the wrinkles of obstinacy, and heaven knows how many other wrinkles. Pathognomical distortion, which accompanies the practice of vice, will likewise, in consequence of the disease it produces, become more distorted and hateful. Thus may the pathognomical expression of friendship, compassion, sincerity, piety, and other moral beauties, become bodily beauty to such as can perceive and admire these qualities. On this is founded the physiognomy of Gallert, which is the only true part of physiognomy. This is of infinite advantage to virtue, and is comprehended in a few words—virtue beautifies, vice deforms."

The branch therefore hath effect, the root none; the fruit has physiognomy, the tree none; the laugh of self-sufficient vanity may therefore arise from the most humble of hearts, and the appearance of folly from the perfection of wisdom. The wrinkles of hypocrisy, therefore, are not the result of any internal power or weakness. The author will always fix our attention on the dial-plate, and will never speak of the power of the watch itself. But take away the dial-plate, and still the hand will go. Take away those pathognomical traits which dissimulation sometimes can effect, and the internal power of impulse will remain. How contradictory therefore is it to say, the traits of folly are there, but not the character of folly; the drop of water is visible, but the fountain, the ocean, is not!

Again. It is certainly incongruous to say, "There is pathognomy, but this is as unnecessary (to be written) as an act of love. It chiefly consists in the motion of the muscles of the countenance and the eyes, and is learned by all men. To teach this would be like an attempt to number the sands of the sea!"

Yet the author in the very next page, with great acuteness begins to teach pathognomy, by explaining twelve of the countenances of Chodowiecki, in which how much is there included of the science of physiognomy!

Give me now leave, my worthy antagonist—yet no longer antagonist, but friend convinced by truth, and the love of truth—I say, give me leave to transcribe, in one continued quotation, some of your excellent thoughts and remarks from your essay, and elucidations on the countenances of Chodowiecki, part of which have been already cited in this fragment, and part not. I am convinced they will be agreeable to my readers.

"Our judgment concerning countenances frequently acquires certainty, not from physiognomical nor pathognomical signs, but from the traces of recent actions, which men cannot shake off. Debauchery, avarice, beggary, have each their livery, by which they are as well known as the soldier by his uniform, or the chimney-sweeper by his sooty jacket. The addition of a trifling expletive in discourse will betray the badness of education; and the manner of putting on the hat what is the company we keep, and what the degree of our folly."

Suffer me here to add, Shall not then the whole form of man discover any thing of his talents and dispositions? Can the most milky candour here forget the straining at a gnat and swallowing a camel?

" Maniacs will often not be known to be disordered in their senses, if not in action. More will often be discovered concerning what a man really is, by his dress, behaviour, and mode of paying his compliments, at his first visit and introduction, in a single quarter of an hour, than in all the time he shall remain. Cleanliness and simplicity of manner will often conceal passions.

" No satisfactory conclusions can often be drawn from the countenances of the most dangerous men. Their thoughts are all concealed under an appearance of melancholy. Whoever has not remarked this, is unacquainted with mankind. The heart of the vicious man is always less easy to be read the better his education has been, the more ambition he has, and the better the company he has been accustomed to keep.

" Cowardice and vanity, governed by an inclination to pleasure and indolence, are not marked with strength equivalent to the mischief they occasion; while, on the contrary, fortitude in defence of justice, against all opponents whatever, be their rank and influence what it may, and the conscious feeling of real self-worth, often look very dangerous, especially when unaccompanied by a smiling mouth.

" Specious as the objections brought by the sophistry

of the sensual may be, it is notwithstanding certain, that there is no possible durable beauty wiihout virtue, and the most hateful deformity may, by the aid of virtue, acquire irresistible charms. Examples of such perfection, among persons of both sexes, I own are uucommon, but not more so than heavenly sincerity, modest compliance without self-degradation, universal philanthropy without busy intrusion, a love of order without being minute, or neatness without foppery, which are the virtues that produce such irresistible charms.

"Vice, in like manner, in persons yielding to its influence, may highly deform; especially when, in consequence of bad education, and want of knowledge of the traits of moral beauty, or of will to assume them, the vicious may find no day, no hour, in which to repair the depredations of vice.

"Where is the person who will not listen to the mouth, in which no trait, no shade of falsehood, is discoverable? Let it preach the experience of what wisdom, what science it may, comfort will ever be the harbinger of such a physician, and confidence hasten to welcome his approach.

"One of the most hateful objects in the creation, says a certain writer, is a vicious and deformed old woman. We may also say that the virtuous matron, in whose countenance goodness and the ardour of benevolence are conspicuous, is an object most worthy our reverence. Age never deforms the countenance when the mind dares appear unmasked; it only wears off the fresh varnish, under which coquetry, vanity, and vice were concealed. Wherever age is exceedingly deformed, the same deformity would have been visible in youth to the attentive observer.

" This is no difficult matter, and were men to act from conviction, instead of flattering themselves with the hope of fortunate accidents, happy marriages would be more frequent; and, as Shakspeare says, the bonds which should unite hearts would not so often strangle temporal happiness."

This certainly is the language of the heart. Oh! that I could have written my fragments in company with such an observer! Who could have rendered greater services to physiognomy than the man who, with the genius of a mathematician, possesses so accurate a spirit of observation?

CHAPTER XXXII.

Description of Plate V.

Number 1.

WILLIAM HONDIUS, a Dutch engraver, after Vandyck. We here see mild, languid, slow industry, with enterprising, daring, conscious heroism. This forehead is rounded, not indeed common nor ignoble. The eyebrows are curved, the eyes languid and sinking, and the whole countenance oval, ductile, and maidenly.

Number 2.

This head, if not stupid, is at least common; if not rude, clumsy. I grant it is a caricature; yet, however, there is something sharp and fine in the eye and mouth, which a connoisseur will discover.

Number 3.

This is manifestly a Turk, by the arching and position of the forehead, the hind part of the head, the eyebrows,

No 1.
No 2.
No 3.
No 4.
No 5.
No 6

and particularly the nose. The aspect is that of observation, with a degree of curiosity: the open mouth denotes remarking, with some reflection.

Number 4.

It must be a depraved taste which can call this graceful, and therefore it must be far from majestic. I should neither wish a wife, mother, sister, friend, relation, nor goddess, to possess a countenance so cold, insipid, affected, stony, unimpassioned, or so perfectly a statue.

Number 5.

The strong grimace of an important madman, who distorts himself without meaning. In the eye is neither attention, fury, littleness, nor greatness.

Number 6.

The eyes in this head are benevolently stupid. Wherever so much white is seen as in the left eye, if in company with such a mouth, there is seldom much wisdom.

CHAPTER XXXIII.

General Remarks on Women.

It may be necessary for me to say, that I am but little acquainted with the female part of the human race. Any man of the world must know more of them than I can pretend to know. My opportunities of seeing them at the theatre, at balls, or at the card-table, where they best may be studied, have been exceedingly few. In my youth I almost avoided women, and was never in love.

Perhaps I ought, for this very reason, to have left this very important part of physiognomy to one much better informed, having myself so little knowledge of the fair sex. Yet might not such neglect have been dangerous? Might another have treated the subject in a manner which I could wish? or, would he have said the little I have to say, and which, though little, I esteem to be necessary and important?

I cannot help shuddering when I think how excessively, how contrary to my intention, the study of physiognomy may be abused when applied to women. Physiognomy will perhaps fare no better than philosophy, poetry, physic, or whatever may be termed art or science. A little philosophy leads to atheism, and much to Christianity. Thus must it be with physiognomy; but I will not be discouraged; the half precedes the whole. We learn to walk by falling, and shall we forbear to walk lest we should fall?

I can with certainty say, that true pure physiognomical sensation, in respect to the female sex, best can season and improve life, and is the most effectual preservative against the degradation of ourselves and others.

Best can season and improve human life.—What better can temper manly rudeness, or strengthen and support the weakness of man, what so soon can assuage the rapid blaze of wrath, what more charm masculine power, what so quickly dissipate peevishness and ill-temper, what so well can while away the insipid tedious hours of life, as the near and affectionate look of a noble, beautiful woman? What is so strong as her soft delicate hand? What so persuasive as her tears restrained? Who but beholding her must cease to sin? How can the spirit of God act more omnipotently upon the heart, than by the extending and increasing physiognomical sensation for such an eloquent countenance? What so well can season daily insipidity? I scarcely can conceive a gift of more paternal and divine benevolence.

This has sweetened every bitter of my life; this alone has supported me under the most corroding cares, when the sorrows of a bursting heart wanted vent, my eyes swam in tears, and my spirit groaned with anguish. Then, when men have daily asked, "Where is now thy God?" when they rejected the sympathy, the affection of my soul, with rude contemptuous scorn; when acts of honest simplicity were calumniated, and the sacred impulse of conscious truth was ridiculed, hissed at, and despised; in those burning moments, when the world afforded no comfort, even then did the Almighty open mine eyes—even then did he give me an unfailing source of joy, contained in a gentle, tender, but internally firm, female mind; an aspect like. that of unpractised, cloistered virginity, which felt and was able to efface each emotion, each passion in the most concealed feature of her husband's countenance, and who by those means, without any thing of what the world calls beauty, shone forth beauteous as an angel. Can there be a more noble or important practice than that of physiognomical sensation for beauties so captivating, so excellent as these?

This physiognomical sensation is the most effectual preservative against the degradation of ourselves and others. —What can more readily discover the boundary between appetite and affection, or cunning under the mask of sensibility? What sooner can distinguish desire from love, or love from friendship? What can more reverently, internally, and profoundly feel the sanctity of innocence, the divinity of maiden purity, or sooner detect coquetry

unblessed, with wiles affecting every look of modesty ?
How often will such a physiognomist turn contemptuous
from the beauties most adored, from the wretched pride
of their silence, their measured affectation of speech, the
insipidity of their eyes, arrogantly overlooking misery
and poverty; their authoritative nose, their languid,
unmeaning lips, relaxed by contempt, blue with envy,
and half-bitten through by artifice and malice! The
obviousness of these and many others will preserve him
who can see, from the dangerous charms of their shame-
less bosoms! How fully convinced is the man of pure
physiognomical sensation, that he cannot be more
degraded than by suffering himself to be ensnared by
such a countenance! Be this one proof among a
thousand.

But if a noble, spotless maiden but appear; all
innocence, and all soul; all love, and of love all worthy,
which must as suddenly be felt as she manifestly feels;
if in her large arched forehead all the capacity of immeasurable intelligence which wisdom can communicate,
be visible; if her compressed but not frowning eyebrows
speak an unexplored mine of understanding, or her gentle
outlined or sharpened nose, refined taste, with sympathetic goodness of heart, which flows through the clear
teeth over her pure and efficient lips; if she breathe
humility and complacency; if condescension and mildness be in each motion of her mouth, dignified wisdom
in each tone of her voice; if her eyes, neither too open
nor too close, but looking straight forward, or gently
turned, speak the soul that seeks a sisterly embrace;
is she be superior to all the powers of description; if all
the glories of her angelic form be imbibed like the mild
and golden rays of an autumnal evening sun; may not

then this so highly-prized, physiognomical sensation be a destructive snare or sin, or both?

"If thine eye be single, thy whole body shall be full of light, as when the bright shining of a candle doth give thee. light." And what is physiognomical sensation but this singleness of eye? The soul is not to be seen without the body, but in the body; and the more it is thus seen, the more sacred to thee will the body be. What! man, having this sensation, which God has bestowed, wouldst thou violate the sanctuary of God? Wouldst thou degrade, defame, debilitate, and deprive it of sensibility? Shall he, whom a good or great countenance does not inspire with reverence and love, incapable of offence, speak of physiognomical sensation; of that which is the revelation of the spirit? Nothing maintains chastity so entire, nothing so truly preserves the thoughts from brutal passion, nothing so reciprocally exalts souls, as when they are mutually held in sacred purity. The contemplation of power awakens reverence, and the picture of love inspires love; not selfish gratification, but that pure passion with which spirits of heaven embrace.

CHAPTER XXXIV.

General Remarks on Male and Female.—A Word on the Physiognomical Relation of the Sexes.

GENERALLY speaking, how much more. pure, tender, delicate, irritable, affectionate, flexible, and patient, is woman than man! The primary matter of which they are constituted appears to be more flexible, irritable, and elastic than that of man. They are formed to maternal mildness and affection. All their organs are tender, yielding, easily wounded, sensible, and receptible.

Among a thousand females there is scarcely one without the generic feminine signs, the flexible, the circular, and the irritable. They are the counterpart of man, taken out of man, to be subject to man; to comfort him like angels, and to lighten his cares. "She shall be safe in child-bearing, if they continue in faith, and charity, and holiness, with sobriety."—(1 Tim. ii. 15.)

This tenderness and sensibility, this light texture of their fibres and organs, this volatility of feeling, render them so easy to conduct and to tempt; so ready of submission to the enterprise and power of the man; but more powerful through the aid of their charms than man, with all his strength. The man was not first tempted, but the woman, afterwards the man by the woman. And not only easily to be tempted, she is capable of being formed to the purest, noblest, most seraphic virtue; to every thing which can deserve praise or affection.

Truly sensible of purity, beauty, and symmetry, she does not always take time to reflect on internal life, internal death, internal corruption. "The woman saw that the tree was good for food, and that it was pleasant to the eyes, and a tree to be desired to make one wise, and she took of the fruit thereof."

The female thinks not profoundly; profound thought is the power of the man. Women feel more: sensibility is the power of women. They often rule more effectually, more sovereignly, than man. They rule with tender looks, tears, and sighs, but not with passion and threats; for if they so rule, they are no longer women, but abortions.

They are capable of the sweetest sensibility, the most profound emotion, the utmost humility, and the excess of enthusiasm. In the countenance are the signs of sanctity and inviolability, which every feeling man honours, and the effects of which are often miraculous. Therefore, by the irritability of their nerves, their incapacity for deep inquiry and firm decision, they may easily, from their extreme sensibility, become the most irreclaimable, the most rapturous enthusiasts.

The love of woman, strong and rooted as it is, is very changeable; their hatred almost incurable, and only to be effaced by continued and artful flattery. Men are most profound, women are more sublime. Men most embrace the whole; women remark individually, and take more delight in selecting the minutiæ which form the whole. Man hears the bursting thunders, views the destructive bolt with serene aspect, and stands erect amidst the fearful majesty of the streaming clouds. Woman trembles at the lightning and the voice of distant thunder, and shrinks into herself, or sinks into the arms of man.

A ray of light is singly received by man; woman delights to view it through a prism, in all its dazzling colours. She contemplates the rainbow as the promise of peace; he extends his inquiring eye over the whole horizon.

Woman laughs, man smiles; woman weeps, man remains silent. Woman is in anguish when man weeps, and in despair when man is in anguish; yet has she often more faith than man. Without religion, man is a diseased creature, who would persuade himself he is well, and needs not a physician : but woman, without religion, is raging and monstrous A woman with a

beard is not so disgusting as a woman who acts the free-
thinker; her sex is formed to pity and religion. To
them Christ first appeared; but he was obliged to
prevent them from too ardently and too hastily embracing him—*Touch me not.* They are prompt to receive
and seize novelty, and become its enthusiasts.

In the presence and proximity of him they love, the
whole world is forgotten. They sink into the most
incurable melancholy, as they rise to the most
enraptured heights.

There is more imagination in male sensation, in the
female more heart. When communicative, they are
more communicative than man; when secret, more
secret. In general they are more patient, long-suffering,
credulous, benevolent, and modest.

Woman is not a foundation on which to build. She
is the gold, silver, precious stones, wood, hay, stubble,
(1 Cor. iii. 12;) the materials for building on the male
foundation. She is the leaven, or, more expressively,
the oil to the vinegar of man; the second part to the
book of man. Man singly is but half a man, at least
but half human; a king without a kingdom. Woman,
who feels properly what she is, whether still or in
motion, rests upon the man; nor is man what he may
and ought to be but in conjunction with woman.
Therefore " It is not good that man should be alone, but
that he should leave father and mother, and cleave to
his wife, and that they two shall be one flesh."

A Word on the Physiognomical Relation of the Sexes.

Man is the most firm, woman the most flexible.
Man is the straightest, woman the most bending.
Man stands steadfast, woman gently retreats.

Man surveys and observes, woman glances and feels.

Man is serious, woman is gay.

Man is the tallest and broadest, woman the smallest and weakest.

Man is rough and hard, woman is smooth and soft.

Man is brown, woman is fair.

Man is wrinkly, woman is not.

The hair of man is strong and short, of woman more long and pliant.

The eyebrows of man are compressed, of woman less frowning.

Man has most convex lines, woman most concave.

Man has most straight lines, woman most curved.

The countenance of man, taken in profile, is not so often perpendicular as that of the woman.

Man is the most angular, woman most round.

CHAPTER XXXV.

On the Physiognomy of Youth.

Extracts from Zimmerman's Life of Haller.

"The first years of the youth include the history of the man. They develop the qualities of the soul, the materials of future conduct, and the true features of temperament. In riper years dissimulation prevails, or, at least, that modification of our thoughts which is the consequence of experience and knowledge.

"The characteristics of the passions, which are undeniably discovered to us by the peculiar art denominated physiognomy, are effaced in the countenance by age; while, on the contrary, their true signs are visible in youth. The original materials of man are unchangeable; he is drawn in colours that have no deceit. The boy is the work of nature, the man of art."

My worthy Zimmerman, how much of the true, how much of the false, at least of the indefinite, is there in this passage! According to my conception, I see the clay, the mass, in the youthful countenance; but not the form of the future man. There are passions and powers of youth, and passions and powers of age. These often are contradictory in the same man, yet are they contained one within the other. Time produces the expression of latent traits. A man is but a boy seen through a magnifying-glass; I always, therefore, perceive more in the countenance of a man than of a boy. Dissimulation may indeed conceal the moral materials, but not alter their form. The growth of powers and passions imparts, to the first undefined sketch of what is called a boy's countenance, the firm traits, shading, and colouring of manhood.

There are youthful countenances which declare whether they ever shall, or shall not, ripen into man. This they declare, but they only declare it to the great physiognomist. I will acknowledge when, which seldom happens, the form of the head is beautiful, conspicuous, proportionate, greatly featured, well defined, and not too feebly coloured, it will be difficult that the result should be common or vulgar. I likewise know that where the form is distorted, especially when it is transverse, extended, undefined, or too harshly defined, much can rarely be expected. But how much do the forms of youthful countenances change, even in the system of the bones !

A great deal has been said of the openness, undegene-

racy, simplicity, and ingenuousness of a childish and
youthful countenance. It may be so ; but, for my own
part, I must own I am not so fortunate as to be able
to read a youthful countenance with the same degree of
quickness and precision, however small that degree, as
one that is manly. The more I converse with and
consider children, the more difficult do I find it to
pronounce, with certainty, concerning their character.
Not that I do not meet countenances, among children
and boys, most strikingly and positively significant ; yet
seldom is the great outline of the youth so definite as for
us to be able to read in it the man. The most remarkably advantageous young countenances may easily,
through accident, terror, hurt, or severity in parents or
tutors, be internally injured, without any apparent injury
to the whole. The beautiful, the eloquent form, the
firm forehead, the deep sharp eye, the cheerful, open,
free, quick-moving mouth remain ; there will only be a
drop of troubled water in what else appears so clear ; only
an uncommon, scarcely remarkable, perhaps convulsive
motion of the mouth. Thus is hope overthrown, and
beauty rendered indistinct.

As simplicity is the soil of variety, so is innocence for
the products of vice. Simplicity, not of a youth, but of
a child, in thee the Omniscient only views the progress
of sleeping passion ; the gentle wrinkles of youth, the
deep of manhood, and the manifold and relaxed of age.
Oh ! how different was my infantine countenance to the
present, in form and speech ! But, as transgression
follows innocence, so doth virtue transgression.

Doth the vessel say to the potter, " Wherefore hast
thou made me thus ?—*I am little, but I am I.*" He
who created me, did not create me to be a child, but a

man. Wherefore should I ruminate on the pleasures of childhood, unburthened with cares? I am what I am. I will forget the past, nor weep that I am no longer a child, when I contemplate children in all their loveliness. To join the powers of man with the simplicity of the child, is the height of all my hopes. God grant they may be accomplished!

CHAPTER XXXVI.

Physiognomical Extracts from an Essay inserted in the Deutschen Museum, a German Journal or Review.

FROM this essay I shall extract only select thoughts, and none but such as I suppose importantly true, false, or ill defined.

1.

"Men with arched and pointed noses are said to be witty, and that the blunt noses are not so."

A more accurate definition is necessary, which, without drawing, is almost impossible. Is it meant by arched noses, arched in length or in breadth? How arched? This is almost as indeterminate as when we speak of arched foreheads. All foreheads are arched. Innumerable noses are arched, the most witty and the most stupid. Where is the highest point of arching? Where does it begin? What is its extent? What is its strength?

It must be allowed that people with tender, thin, sharply-defined, angular noses, pointed below, and something inclined towards the lip, are witty, when no other features contradict these tokens; but that people with blunt noses are not so, is not entirely true. It can only be said of certain blunt noses; for there are others of this kind extremely witty, though their wit is certainly of a different kind to that of the pointed nose.

2

" It is asked, (supposing for a moment that the arched and the blunt nose denote the presence or absence of wit,) Is the arched nose the mere sign that a man is witty, which supposes his wit to originate in some occult cause, or is the nose itself the cause of wit ? "

I answer, sign, cause, and effect combined. Sign ; for it betokens the wit, and is an involuntary expression of wit. Cause ; at least cause that the wit is not greater, less, or of a different quality, boundary cause. Effect ; produced by the quantity, measure, or activity of the mind, which suffers not the nose to alter its form, to be greater or less. We are not only to consider the form as form, but the matter of which it is moulded, the conformability of which is determined by the nature and ingredients of this matter, which is probably the origin of the form.

True indeed it is, that there are blunt noses which are incapable of receiving a certain quantity of wit ; therefore it may be said, with more subtlety than philosophy, they form an insuperable barrier.

3.

" The correspondence of external figures with internal qualities is not the consequence of external circumstances, but rather of physical combination. They are related like cause and effect; or, in other words, physiognomy is not the mere image of internal man, but the efficient

cause. The form and arrangement of the muscles determine the mode of thought and sensibility of the man."

I add, these are also determined by the mind of man.

4.

"A broad conspicuous forehead is said to denote
penetration. This is natural. The muscle of the forehead is necessary to deep thought. If it be narrow and
contracted, it cannot render the same service as if spread
out like a sail."

I shall here, without contradicting the general proposition of the author, more definitely add—It is, if you
please, generally true, that the more brain the more
mind and capacity. The most stupid animals are those
with least brain, and those with most the wisest. Man,
generally wiser, has more brain than other animals;
and it appears just to conclude from analogy, that wise
men have more brain than the foolish. But accurate
observation teaches, that this proposition, to be true,
requires much definition and limitation.

Where the matter and form of the brain are similar,
there the greater space for the residence of the brain is,
certainly the sign, cause, and effect of more and deeper
impression; therefore, *cæteris paribus*, a larger quantity
of brain, and consequently a spacious forehead, is more
intelligent than the reverse. But as we frequently live
more conveniently in a small well-contrived chamber
than in more magnificent apartments, so do we find that
in many small, short foreheads, with less, or apparently
less brain than others, the wise mind resides at its ease.

I have known many short, oblique, straight-lined
(when compared with others apparently arched, or really
well arched) foreheads, which were much wiser, more

intelligent, and penetrating, than the most broad and conspicuous; many of which latter I have seen in extremely weak men. It seems to me, indeed, a much more general proposition, that short compressed foreheads are wise and understanding; though this, likewise, without being more accurately defined, is far from being generally true.

But it is true that large spacious foreheads, which, if I do not mistake, Galen, and after him Huart, have supposed the most propitious to deep thinking, which form a half sphere, are usually the most stupid. The more any forehead (I do not speak of the whole skull) approaches a semi-spherical form, the more is it weak, effeminate, and incapable of reflection, and this I speak from repeated experience.

The more straight lines a forehead has, the less capacious it must be; for the more it is arched the more must it be roomy, and the more straight lines it has the more must it be contracted. This greater quantity of straight lines, when the forehead is not flat like a board, for such flatness takes away all understanding, denotes an increase of judgment, but a diminution of sensibility. There undoubtedly are, however, broad capacious foreheads, without straight lines, particularly adapted to profound thinking; but these are conspicuous by their oblique outlines.

5.

What the author has said concerning enthusiasts requires much greater precision before it ought to be adopted as true.

"Enthusiasts are said commonly to have flat perpendicular foreheads."

Oval, cylindrical, or pointed at top, should have been said of those enthusiasts who are calm, cold-blooded, and always continue the same. Other enthusiasts, that is to say, such as are subject to a variety of sensation, illusion, and sensual experience, seldom have cylindrical or sugar-loaf heads. The latter when enthusiasts, heat their imagination concerning words and types, the signification of which they do not understand, and are philosophical, unpoetical enthusiasts. Enthusiasts of imagination or of sensibility seldom have flat forms of the countenance.

6.

"Obstinate, like enthusiastic, persons have perpendicular foreheads."

The perpendicular always denotes coldness, inactivity, narrowness; hence firmness, fortitude, pertinacity, obstinacy, and enthusiasm may be there. Absolute perpendicularity and absolute folly are the same.

7.

"Such disposition of mind is accompanied by a certain appearance or motion of the muscles; consequently the appearance of man, which is natural to, and ever present with him, will be accompanied by, and denote his natural disposition of mind. Countenances are so formed originally, that to one this, and to another that appearance is the easiest. It is absolutely impossible for folly to assume the appearance of wisdom, otherwise it would no longer be folly. The worthy man cannot assume the appearance of dishonesty, or he would be dishonest."

This is all excellent, the last excepted. No man is so

good as not, under certain circumstances, to be liable to become dishonest. He is so organized that he may be so overtaken by the pleasure of stealing, when accompanied by the temptation. The possibility of the appearance must be there, as well as the possibility of the act. He must also be able to assume the appearance of dishonesty when he observes it in a thief, without necessarily becoming a thief. The possibility of assuming the appearance of goodness is, in my opinion, very different. The appearance of vice is always more easily assumed by the virtuous, than the appearance of virtue by the vicious; as it is evidently much easier to become bad when we are good, than good when we are bad. Understanding, sensibility, talents, genius, virtue, or religion, may with much greater facility be lost than acquired. The bést may descend as low as · they please, but the worst cannot ascend to the height they might wish. The wise man may physically, without a miracle, become a fool, and the most virtuous vicious; but the idiot-born cannot, without a miracle, become a philosopher, nor the distorted villain noble and pure of heart. The most beautiful complexion may become jaundiced, may be lost; but the Negro cannot be washed white. I shall not become a Negro because, to imitate him, I blacken my face, nor a thief because I assume the appearance of one.

8.

"It is the business of a physiognomist to inquire what is the appearance the countenance can most easily assume, and he will thence learn what is the disposition of mind; not that physiognomy is therefore an easy

o

science. On the contrary, this rather shows how much ability, imagination, and genius, are necessary to the physiognomist. Attention must not only be paid to what is visible, but what would be visible under various other circumstances."

This is excellent; and I add, that as a physician can presage what alteration of colour, appearance, or form, shall be the consequence of a known disease, of the existence of which he is certain, so can the accurate physiognomist what appearances or expressions are easy or difficult to each kind of muscle and form of forehead, what action is or is not permitted, and what wrinkles may or may not take place, under any given circumstances.

9.

"When a learner draws a countenance, we shall commonly find it is foolish, and never malicious, satirical, and the like. May not the essence of a foolish countenance hence be abstracted? Certainly; for what is the cause of this appearance? The learner is incapable of preserving proportion, and the strokes are unconnected. What is the stupid countenance? It is one, the parts of which are defectively connected, and the muscles improperly formed and arranged. Thought and sensation, therefore, of which these are the inseparable instruments, must be alike feeble and dormant.

10.

"There is another substance in the body, exclusive of the muscles; that is to say, the skull, or bones in general, to which the physiognomist attends. The position of the muscles depends on these. How might the muscle of the forehead have the position proper for thought, if the forehead bones, over which it is extended, had not the necessary arch and superficies ? The figure of the skull, therefore, defines the figure and position of the muscles which define thought and sensation."

11.

"The hair affords us the same observation, as from the parts and position of the hair conclusions may be drawn. Why has the Negro woolly hair ? Tho thickness of the skin prevents the escape of certain of the particles of perspiration, and these render the skin opaque and black. Hence the hair shoots with difficulty, and scarcely has it penetrated before it curls, and its growth ceases. The hair spreads according to the form of the skull and the position of the muscles, and gives occasion to the physiognomist to draw conclusions from the hair to the position of the muscles, and to deduce other consequences."

It is clearly my opinion that our author is in the right road. He is the first who, to my knowledge, has perceived and felt the totality, the combination, the uniformity, of the various parts of the human body. What he has affirmed, especially concerning the hair, that we may from that make deductions concerning the nature of the body, and still farther of the mind, the least accurate observer may convince himself is truth, by daily experience. White, tender, clear, weak hair, always denotes weak, delicate, irritable, or rather a timid and easily oppressed organization. The black and curly will never be found on the delicate, tender, medullary head.

As is the hair, so the muscles; as the muscles, so the nerves; as the nerves, so the bones: their powers are mutual, and the powers of the mind to act, suffer, receive, and give, proportionate. Least irritability always accompanies short, hard, curly, black hair, and the most the flaxen and the tender; that is to say, irritability without elasticity. The one is oppressive without elasticity, and the other oppressed without resistance.

Much hair, much fat; therefore, no part of the human body is more conspicuously covered with hair than the head and armpits. From the elasticity of the hair, deductions may with certainty be made to the elasticity of the character. The hair naturally betokens moisture, and may properly determine the quantity of moisture. The inhabitants of cold countries have hair more white, and, on the contrary, those of hot countries, black. Lionel Wafer observes, that the inhabitants of the isthmus of Darien have milkwhite hair. Few, if any, have green hair, except those who work in copper mines. We seldom find white hair betokening dishonesty, but often dark brown or black, with light-coloured eyebrows. Women have longer hair than men. Men with long hair are always rather effeminate than manly. Dark hair is harsher than light, as is the hair of a man than that of a boy.

12.

"As all depends on the quality of the muscles, it is evident that in these muscles which are employed for certain modes of thought and sensation, ought to be sought the expression of similar thoughts and sensations."

The search should not be neglected, though perhaps it will be difficult to find them; and they certainly will there be defined with greater difficulty than in the forehead.

13.

"The most important instrument to the abstract thinker is the muscle of the forehead; for which reason we always seek for abstract thought in the forehead."

Rather near and between the eyebrows. It is of consequence to remark the particular moment when the thinker is listening, or when he is preparing some acute answer. Seize the moment, and another of the important tokens of physiognomy is obtained.

14.

"Among people who do not abstract, and whose powers of mind are all in action, men of wit, exquisite taste, and genius, all the muscles must be advantageously formed and arranged. Expression therefore, in such, must be sought in the whole countenance.

Yet may it be found in the forehead alone, which is less sharp, straight-lined, perpendicular, and forked. The skin is less rigid, more easily moved, more flexible.

15.

"How laborious has been the trouble to convince people that physiognomy is only generally useful!"

It is at this very moment disputed by men of the strongest minds. How long shall it continue so to be? Yet I should suppose that he who curses the sun while exposed to its scorching rays, would, when in the shade, acknowledge its universal utility.

"How afflicting is it to hear, from persons of the greatest learning, and who might be expected to enlarge the boundaries of human understanding, the most superficial judgments! How much is that great era to be wished, when the knowledge of man shall become a part of natural history; when psychology, physiology, and physiognomy, shall go hand in hand, and lead us towards the confines of more general, more sublime illumination!"

CHAPTER XXXVII.

Extracts from Maximus Tyrius.

"As the soul of man is the nearest approach to the Deity, it was not proper that God should clothe that which most resembled himself in dishonourable garments; but with a body befitting a mortal mind, and endowed with a proper capability of motion. This is the only body on earth that stands erect. It is magnificent, superb, and formed according to the best proportion of its most delicate parts. Its stature is not terrific, nor is its strength formidable. The coldness of its juices occasions it not to creep, nor their heat to fly. Man eats not raw flesh from the savageness of his nature, nor does he graze like the ox; but he is framed and adapted for the execution of his functions. To the wicked he is formidable, mild and friendly to the good. By nature he walks the earth, swims by art, and flies by imagination. He tills the earth, and enjoys its fruits. His complexion is beautiful, his limbs firm, his countenance is comely, and beard ornamental. By imitating his body, the Greeks have thought proper to honour their deities."

Why am I not able to speak with sufficient force?

Oh! that I could find faith enough with my readers, to convince them how frequently my soul seems exalted above itself, while I contemplate the unspeakably miraculous nature of the human body! Oh! that all the languages of the earth would lend me words, that I might turn the thoughts of men, not only to the contemplation of others, but, by the aid of these, to the contemplation of themselves! No anti-physiognomist can more despise my work than I myself shall, if I am unable to accomplish this purpose. How might I conscientiously write such a work, were not such my views? If this be not impulse, no writer has impulse. I cannot behold the smallest trait, nor the inflection of any outline, without reading wisdom and benevolence, or without waking as if from a sweet dream into rapturous and actual existence, and congratulating myself that I also am a man.

In each, the smallest outline of the human body, and how much more in all together—in each member separately, and how much more in the whole body, however old and ruinous the building may appear—how much is there contained of the study of God, the genius of God, the poetry of God? My trembling and agitated breast frequently pants after leisure to look into the revelations of God.

2.

"Imagine to thyself the most translucent water flowing over a surface on which grow beauteous flowers, whose bloom, though beneath, is seen through the pellucid waves; even so it is with the fair flower of the soul, planted in a beauteous body, through which its beauteous bloom is seen. The good formation of a youthful body is no other than the bloom of ripening virtue, and, as I may say, the presage of far higher perfection; for, as before the rising of the sun, the mountain-tops are gilded by his rays, enlivening the pleasing prospects, and promising the full approach of day, so also the future maturity of an illustrious soul shines through the body, and is to the philosopher the pleasing sign of approaching happiness."

CHAPTER XXXVIII.

Extracts from a Manuscript by Th———.

"THE relation between the male and female countenance is similar to that between youth and manhood. Our experience, that the deep or scarcely visible outline is in proportion to the depth or shallowness of thought, is one of the many proofs that Nature has impressed such forms upon her creatures as shall testify their qualities. That these forms or signs are legible to the highly perceptive soul is visible in children, who cannot endure the deceitful, the tell-tale, or the revengeful; but run with open arms to the benevolent stranger.

" We may properly divide our remarks on this subject into complexion, lines, and pantomime. That white, generally speaking, is cheerful, and black gloomy and terrific, is the consequence of our love of light, which acts so degenerately, as it were, upon some animals, that they will throw themselves into the fire; and of our abhorrence of darkness. The reason of this our love of light is, that it makes us acquainted with things, provides for the soul hungry after knowledge, and enables us to find what is necessary, and avoid what is dangerous. I only mention this to intimate, that this our love of light originates in our inclination for every thing that is perspicuous. Certain colours are, to certain animals, particularly agreeable or disagreeable."

What is the reason of this? Because they are the expression of something which has a relation to their character, that harmonizes with it or is discordant. Colours are the effect of certain qualities of object and subject; they are therefore characteristic in each, and become more so by the manner in which they are mutually received and repelled. This would be another immense field of inquiry, another ray of the sun of truth. All is physiognomy!

"Our dislike is no less for every thing which is clothed in dark colours; and nature has warned animals, not only against feeding on earth, but also on dark-green plants; for the one is as detrimental as the other. Thus the man of a dark complexion terrifies an infant that is incapable of judging of his character.

"So strikingly significant are the members of the body, that the aspect of the whole attacks our feelings, and induces judgments as sudden as they are just. Thus, to mention two extremes, all will acknowledge at the first aspect the elephant to be the wisest, and the fish the most stupid of creatures.

"The upper part of the countenance, to the root of the nose, is the seat of internal labour, thought, and re-solution; the under, of these in action. Animals with very retreating foreheads have little brain, and the reverse.

"Projecting nose and mouth betoken persuasion, self-confidence, rashness, shamelessness, want of thought, dishonesty, and all such feelings as are assembled in hasty expression."

This is a decision after the manner of the old physiognomists, condemning and indefinite.

"The nose is the seat of derision; its wrinkles contemn. The upper lip, when projecting, speaks arrogance, threats, and want of shame; the parting under lip, ostentation and folly. These signs are confirmed by the manner and attitude of the head when drawn back, tossed, or turned round. The first expresses contempt, during which the nose is active; the latter is a proof of extreme arrogance, during which the projection of the under lip is the strongest.

"The in-drawn lower parts of the countenance, on the contrary, denote discretion, modesty, seriousness, diffidence, and its failings are those of malice and obstinacy."

Not so positive. The projecting chin is much oftener the sign of craft than the retreating. The latter is seldom scheming and enterprising.

"The straight formation of the nose betokens gravity; inbent and crooked, noble thoughts. The flat, pouting upper lip, when it does not close well with the under, signifies timidity; the lips resembling each other, circumspection of speech."

"We may divide the face into two principal kinds. The first is that in which the cheeks present a flat surface, the nose projecting like a hill, and the mouth has the appearance of a sabre wound prolonged on an even surface, while the line of the jawbone has but little inflection. Such a form makes the countenance more broad than long, and exceedingly rude, inexpressive, stupid, and in every sense confined. The principal characteristics are obstinacy and inflexibility.

"The second kind is, when the nose has a sharp ridge,

and the parts on both sides make acute angles with each other. The cheekbones are not seen, consequently the muscular parts between them and the nose are full and prominent. The lips retreat on each side of the mouth, assume or open into an oval, and the jawbones come to a point at the chin."   `

This face denotes a mind more subtle, active, and intelligent.

"The better to explain myself, I must here employ the simile of two ships. The first, a merchant vessel built for deep loading, has a broad bottom, and her ribs long and flat. This resembles the broad, flat countenance. The frigate, built for swift sailing, has a sharp keel or bottom, her ribs forming acute angles. Such is the second countenance. Of these two extremes, the first presents to me the image of the meanest, most contracted, self-love; the second of the most zealous, the noblest philanthropy.

"I am sensible that nature does not delight in extremes. Still the understanding must take its departure from these as from a lighthouse, especially when sailing in unknown seas. The defects and excesses which are in all works of nature will then be discovered, and one or both the boundaries ascertained.

"If we proceed to a farther examination and application of the above hypothesis, it will perhaps extend through all nature. A broad countenance is accompanied by a short neck, broad shoulders and back, and their known character is selfishness and obtuse sensation. The long small countenance has a long neck, small or low shoulders, and small back. From such I should expect more justice, disinterestedness, and a general superiority of social feelings.

"The features and character of men are essentially altered by education, situation, intercourse, and incidents; therefore we are justified in maintaining, that physiognomy cannot look back to the origin of the features, nor presage the changes of futurity; but from the countenance only, abstracted from all external accidents by which it may be affected, it may read what any given man may be, with the following addition at most: such shall be the empire of reason, or such the power of sensuality. This man is too stubborn to be instructed: that so flexible, he may be led to good or ill.

"From this formation we may in part explain why so many men appear to be born for certain situations, although they may have rather been placed in them by accident than by choice. Why the prince, the nobleman, the overseer of the poor, have a lordly, a stern, or a pedantic manner; why the subject, the servant, the slave, are pusillanimous and spiritless; or the courtesan affected, constrained, or insipid. The constant influence of circumstances on the mind far exceeds the influence of nature." Far the contrary.

"Although it is certain that *innate* servility is very distinct from the servility of one whom misfortune has rendered a servant; like as he whom chance has made a ruler over his brother, is very different from one who is by nature superior to vulgar souls."

There is no such thing as *innate* servility. It is true that, under certain circumstances, some are much more disposed than others to become servile.

"The unfeeling mind of the slave has vacuity more complete, or, if a master, more self-complacency and arrogance, in the open mouth, the projecting lip, and the turned-up nose. The noble mind rules by the comprehensive aspect, while in the closed lips moderation is expressed. He will serve with sullenness, with downcast eye, and his shut mouth will disdain to complain.

"These causes will undoubtedly make durable impressions, so will the adventitious occasion transitory ones, while their power remains. The latter are more apparent than the signs of the countenance at rest, but may be well defined by the principal characteristics of the agitated features; and, by comparison with countenances subject to similar agitations, the nature of the mind may be fully displayed. Anger in the unreasonable, ridiculously struggles; in the self-conceited, it is fearful rage; in the noble-minded it yields, and brings opponents to shame; in the benevolent, it has a mixture of compassion for the offender, moving him to repentance.

"The affliction of the ignorant is outrageous, and of the vain ridiculous; of the compassionate, abundance in tears and communicative; of the resolute serious, internal, the muscles of the cheeks scarcely drawn upwards, the forehead little wrinkled.

"Violent and eager is the love of the ignorant; of the vain, disgusting, which is seen in the sparkling eyes, and the forced smile of the forked cheeks, and the indrawn mouth; of the tender, languishing, with the mouth contracted to entreat; of the man of sense, serious, steadfastly surveying the object, the forehead open, and the mouth prepared to plead.

"On the whole, the sensations of a man of fortitude are restrained, while those of the ignorant degenerate into grimace. The latter, therefore, are not the proper study of the artist, though they are of the physiognomist and the moral teacher, that youth may be warned against too strong an expression of the emotions of the mind, and of their ridiculous effects.

"In this manner do the communicative and moving sensations of the benevolent inspire reverence ; but those of the vicious, fear, hatred, or contempt.

"The repetition of passions engraves their signs so deeply, that they resemble the original stamp of nature. Hence certainly may be deduced, that the mind is addicted to such passions. Thus are poetry and the dramatic art highly beneficial, and thus may be seen the advantage of conducting youth to scenes of misery and of death.

"Such a similarity is formed by frequent intercourse between men, that they not only assume a mental likeness, but frequently contract some resemblance of voice and features. Of this I know several examples.

"Each man has his favourite gesture, which might decipher his whole character, might he be observed with sufficient accuracy to be drawn in that precise posture. The collection of such portraits would be excellent for the first studies of the physiognomist, and increase the utility of the fragments of Lavater tenfold.

"A series of drawings of the motions peculiar to individuals, would be of equal utility. The number of them in lively men is great, and they are transitory. In the more sedate, they are less numerous and more grave.

"As a collection of idealized individuals would promote an extensive knowledge of various kinds of men, so would a collection of the motions of a single countenance promote a history of the human heart, and demonstrate what an arrogant, yet pusillanimous thing the unformed heart is, and the perfection it is capable of, from the efforts of reason and experience.

" It would be an excellent school for youth to see Christ teaching in the Temple, asking, Whom seek you? . agonizing in the Garden, expiring on the Cross. Ever the same God-man! Ever displaying, in these various situations, the same miraculous mind, the same steadfast reason, the same gentle benevolence. Cæsar jesting with the pirates when their prisoner, weeping over the head of Pompey, sinking beneath his assassins, and casting an expiring look of affliction and reproach while he exclaims, *Et tu, Brute?* Belshazzar, feasting with his nobles, turning pale at the handwriting on the wall. The tyrant enraged, butchering his slaves, and surrounded by condemned wretches entreating mercy from the uplifted sword.

" Sensation having a relative influence on the voice, must not there be one principal tone or key by which all the others are governed; and will not this be the key in which he speaks when unimpassioned, like as the countenance at rest contains the propensities to all such traits as it is capable of receiving? These keys of voice a good musician with a fine ear should collect, class, and learn to define, so that he might place the key of the voice beside any given countenance, making proper allowances for changes occasioned by the form of the lungs, exclusive of disease. Tall people, with a flatness of breast, have weak voices.

" This idea, which is more difficult to execute than conceive, was inspired by the various tones in which I have heard *yes* and *no* pronounced. The various emotions under which these words are uttered, whether of assurance, decision, joy, grief, ridicule, or laughter, will give birth to tones as various. Yet each man has his peculiar manner, respondent to his character, of saying *yes, no,* or any other word. It will be open, hesitating, grave, trifling, sympathizing, cold, peevish, mild, fearless, or timid. What a guide for the man of the world, and how do such tones display or betray the mind!

"Since we are taught by experience, that at certain times the man of understanding appears foolish, the courageous cowardly, the benevolent perverse, and the cheerful discontented, we might, by the assistance of these accidental traits, draw an idea of each motion; and this would be a most valuable addition, and an important step in the progress of physiognomy."

CHAPTER XXXIX.

Extracts from Nicolai and Winkelmann.

Extracts from Nicolai.

1.

"THE distorted or disfigured form may originate as well from external as from internal causes; but the consistency of the whole is the consequence of conformity between internal and external causes; for which reason moral goodness is much more visible in the countenance than moral evil."

This is true, those moments excepted when moral evil is in act.

2.

"The end of physiognomy ought to be, not conjectures on individual, but the discovery of general character."

The meaning of which is, the discovery of general signs of powers and sensations, which certainly are useless unless they can be individually applied, since our intercourse is with individuals.

3.

"It would be of great utility to physiognomy were numerous portraits of the same man annually drawn, and the original by that means well known."

It is possible, and perhaps only possible, to procure accurate shades or plaster casts. Minute changes are seldom accurately enough attended to by the painter, for the purpose of physiognomy.

4.

"The most important pursuit of the physiognomist in his researches will ever be, in what manner is a man considered capable of the impressions of sense? Through what kind of perspective does he view the world? What can he give? What receive?

5.

"That very vivacity of imagination, that quickness of conception, without which no man can be a physiognomist, is probably almost inseparable from other qualities, which render the highest caution necessary if the result of his observations is to be applied to living persons."

This I readily grant; but the danger will be much less if he endeavour to employ his quick sensations in determinate signs; if he be able to portray the general tokens of certain powers, sensations, and passions, and if his rapid imagination be only busied to discover and draw resemblances.

P

Extracts from Winkelmann.

1.

" The characteristic of truth is internal sensation, and the designer who would present such natural sensation to his academy, would not obtain a shade of the true, without a peculiar addition of something, which an ordinary and unimpassioned mind cannot read in any model, being ignorant of the action peculiar to each sensation and passion.

" The physiognomist is formed by internal sensation, which, if the designer be not, he will give but the shadow, and only an indefinite and confused shadow, of the true character of nature."

2.

" The forehead and nose of the Greek gods and goddesses form almost a straight line. The heads of famous women on Greek coins have similar profiles, where the fancy might not be indulged in ideal beauties. Hence we may conjecture that this form was as common to the ancient Greeks as the flat nose to the Calmuc, or the small eye to the Chinese. The large eyes of Grecian heads in gems and coins support this conjecture."

This ought not to be absolutely general, and probably was not, since numerous medals show the contrary, though in certain ages and countries such might have been the most common form. Had only one such countenance, however, presented itself to the genius of art, it would have been sufficient for its propagation and continuance. This is less our concern than the signification of such a form. The nearer the approach to the perpendicular, the less is there characteristic of the wise and graceful; and the higher the character of worth and greatness, the more obliquely the lines retreat. The more straight and perpendicular the profile of the forehead and nose is, the more does the profile of the upper part of the head approach a right angle, from which wisdom and beauty will fly with equally rapid steps. In the usual copies of these famous ancient lines of beauty, I generally find the expression of meanness, and, if I dare say, of vague insipidity. I repeat, in the copies; in the Sophonisba of Angelica Kauffman, for instance, where probably the shading under the hair has been neglected, and where the gentle arching of the line apparently were scarcely attainable.

3.

"The line which separates the repletion from the excess of nature, is very small."

Not to be measured by industry or instrument, yet all powerful, as every thing unattainable is.

4.

"A mind as beautiful as was that of Raphael, in an equally beautiful body, is necessary, first to feel, and afterwards to display, in these modern times, the true character of the ancients.

5.

"Constraint is unnatural, and violence disorder."

Where constraint is remarked, there let secret, profound, slowly destructive passion be feared; where violence, there open and quick destroying.

6.

"Greatness will be expressed by the straight and replete, and tenderness by the gently curving."

All greatness has something of straight and replete; but all the straight and replete is not greatness. The straight and replete must be in a certain position, and must have a determinate relation to the horizontal, on which the observer stands to view it.

"It may be proved that no principle of beauty exists in this profile; for the stronger the arching of the nose is, the less does it contain of the beautiful; and if any countenance seen in profile is bad, any search after beauty will there be to no purpose."

The noblest, purest, wisest, most spiritual and benevolent countenance, may be beautiful to the physiognomist, who, in the extended sense of the word beauty, understands all moral expressions of good as beautiful; yet the form may not therefore, accurately speaking, deserve the appellation of beautiful.

7.

"Nothing is more difficult than to demonstrate a self-evident truth."

CHAPTER XL.

Extracts from Aristotle and other Authors concerning Beasts.

THE writings of the great Aristotle on physiognomy appear to me very superficial, useless, and often self-contradictory, especially his general reasoning. Still, however, we sometimes meet an occasional thought which deserves to be selected. The following are some of these :—

"A monster has never been seen which had the form of another creature, and, at the same time, totally different powers of thinking and acting. Thus, for example, the groom judges from the mere appearance of the horse ; the huntsman, from the appearance of the hound. We find no man entirely like a beast, although there are some features in man which remind us of beasts.

" Those who would endeavour to discover the signs of bravery in man, would act wisely to collect all the signs of bravery in animated nature, by which courageous animals are distinguished from others. The physiogno-mist should then examine all such animated beings, which are the reverse of the former with respect to internal character, and, from the comparison of these opposites, the expressions or signs of courage would be manifest.

" As weak hair is a mark of fear, so is strong hair of courage. This observation is applicable not only to men but to beasts. The most fearful of beasts are the deer, the hare, and the sheep, and the hair of these is weaker than that of other beasts. The lion and wild-boar, on the contrary, are the most courageous, which property is conspicuous in their extremely strong hair. The same also may be remarked of birds ; for, in general, those among them which have coarse feathers are courageous, and those that have soft and weak feathers are fearful.

" This may easily be applied to men. The people of the north are generally courageous, and have strong hair ; while those of the west are more fearful, and have more flexible hair.

"Such beasts as are remarkable for their courage simply give their voices vent, without any great constraint, while fearful beasts utter vehement sounds. Compare the lion, ox, the barking dog, and cock, which are courageous, to the deer and the hare. The lion appears to have a more masculine character than any other beast. He has a large mouth, a four-cornered not too bony visage. The upper-jaw does not project, but exactly fits the under; the nose is rather hard than soft, the eyes are neither sunken nor prominent, the forehead is square, and sometimes flattened in the middle.

"Those who have thick and firm lips, with the upper hung over the under, are simple persons, according to the analogy of the ape and monkey."

This is most indeterminately spoken. He would have been much more true and accurate had he said, those whose under-lips are weak, extended, and projecting beyond the upper, are simple people.

"Those who have the tip of the nose hard and firm, love to employ themselves on subjects that give them little trouble, similar to the cow and the ox."

Insupportable! The few men, who have the tip of the nose firm, are the most unwearied in their researches. I shall transcribe no farther. His physiognomical remarks, and his similarities to beasts, are generally unfounded in experience.

Porta, next to Aristotle, has most observed the resemblance between the countenances of men and beasts, and has extended this inquiry the farthest. He, as far as I know, was the first to render this similarity apparent, by placing the countenances of men and beasts beside each other. Nothing can be more true than this fact; and, while we continue to follow nature, and do not endeavour to make such similarities greater than they are, it is a subject that cannot be too accurately examined. But in this respect the fanciful Porta appears to me to have been often misled, and to have found resemblances which the eye of truth never could discover. I could find no resemblance between the hound and Plato, at least from which cool reason could draw any conclusions. It is singular enough that he has also compared the heads of men and birds. He might more effectually have examined the excessive dissimilarity than the very small and almost imperceptible resemblance which can exist. He speaks little concerning the horse, elephant, and monkey, though it is certain that these animals have most resemblance to man.

A generic difference between man and beast is particularly conspicuous in the structure of the bones. The head of man is placed erect on the spinal bone. His whole form is as the foundation pillar for that arch in which heaven should be reflected, supporting that skull by which, like the firmament, it is encircled. This cavity for the brain constitutes the greater part of the head. All our sensations, as I may say, ascend and descend above the jawbone, and collect themselves upon the lips. How does the eye, that most eloquent of organs, stand in need, if not of words, at least of the angry constraint of the cheeks, and all the intervening shades, to express the strong internal sensation of man!

The formation of beasts is directly the reverse of this. The head is only attached to the spine. The brain, the extremity of the spinal marrow, has no greater extent than is necessary for animal life, and the conducting of a creature wholly sensual, and formed but for temporary existence. For although we cannot deny that beasts have the faculty of memory, and act from reflection ; yet the former, as I may say, is the effect of primary sensation, and the latter originates in the constraint of the moment, and the preponderance of this or that object.

We may perceive in the most convincing manner, in the difference of the skull, which defines the character of animals, how the bones determine the form, and denote the properties of the creature.

As the character of animals are distinct, so are their forms, bones, and outlines. From the smallest winged insect to the eagle that soars and gazes at the sun; from the weakest worm impotently crawling beneath our feet, to the elephant or the majestic lion, the gradations of physiognomical expression cannot be mistaken. It would be more than ridiculous to expect from the worm, the butterfly, and the lamb, the power of the rattlesnake, the eagle, and the lion. Were the lion and lamb, for the first time, placed before us, had we never known such animals, never heard their names, still we could not resist the impression of the courage and strength of the one, or of the weakness and sufferance of the other.

Let me ask the question, Which are, in general, the weakest animals, and the most remote from humanity, the most incapable of human ideas and sensations ? Beyond all doubt, those which in their form least resemble man. To prove this, let us, in imagination, consider the various degrees of animal life, from the smallest animalcule to the ape, lion, and elephant; and, the more to simplify and give facility to such comparison, let us only compare head to head; as, for example, the lobster to the elephant, the elephant to the man.

Permit me here just to observe, how worthy would such a work be of the united abilities of a Buffon, a

Kamper, and a Euler, could they be found united, that the forms of heads might be enumerated and described philosophically and mathematically; that it might be demonstrated that universal brutality, in all its various kinds, is circumscribed by a determinate line; and that, among the innumerable lines of brutality, there is not one which is not internally and essentially different from the line of humanity, which is peculiar and unique.

Thoughts of a Friend on Brutal and Human Physiognomy.

"Every brute animal is distinguished from all others by some principal quality. As the make of each is distinct from all others, so also is the character. This principal character is denoted by a peculiar and visible form. Each species of beast has certainly a peculiar character, as it has a peculiar form. May we not hence by analogy infer, that predominant qualities of the mind are certainly expressed by predominant forms of the body, as that the peculiar qualities of a species are expressed in the general form of that species?

"The principal character of the species in animals remain such as it was given by nature; it neither can be obscured by accessory qualities, nor concealed by art. The essential of the character can as little be changed as the peculiarity of the form. May we not therefore, with the greatest degree of certainty, affirm such a form is only expressive of such a character?

"Let us now inquire whether this be applicable to man, and whether the form, which denotes individual character in a beast, is significant of similar character in man, granting that in man, it may continually be more delicate, hidden, and complicated. If, on examination, this question be definitely answered in the affirmative, how much is thereby gained! But it is conspicuously evident that in man the mind is not one character or quality, but a world of qualities interwoven with and obscuring each other. If each quality be expressed by its peculiar from, then must variety of qualities be attended with variety of forms; and these forms, combining and harmonizing together, must become more difficult to select and decipher.

"May not souls differ from each other merely according to their relative connection with bodies? May not souls also have a determinate capacity, proportionate to the form and organization of the body? Hence, each object may make a different impression on each individual; hence one may bear greater burthens and more misfortunes than another. May not the body be considered as a vessel with various compartments, cavities, pipes, into which the soul is poured, and, in consequence of which, motion and sensation begin to act? And thus may not the form of the body define the capacity of the mind?"

My unknown friend, thus far have I followed you. Figurative language is dangerous when discoursing on the soul; yet how can we discourse on it otherwise? I pronounce no judgment, but rely on sensation and experience, not on words and metaphors. What is is, be your language what it will. Whether effects all act from the external to the internal, or the reverse, I know not, cannot, need not know. Experience convinces us that, both in man and beast, power and form are unchangeable, harmonized proportion; but whether the form be determined by the power, or the power by the form, is a question wholly insignificant to the physiognomist.

Observations on some Animals, and particularly of the Horse.

The dog has more forehead above the eyes than most other beasts; but as much as he appears to gain in the forehead he loses in the excess of brutal nose, which has every token of acute scent. Man too, in the act of smelling, elevates the nostrils. The dog is also defective in the distance of the mouth from the nose, and in the meanness or rather nullity of the chin.

Whether the hanging ears of a dog are characteristic of slavish subjection, as Buffon has affirmed, who has written much more reasonably on brute than on human physiognomy, I cannot determine to my own satisfaction.

The camel and the dromedary are a mixture of the horse, sheep, and ass, without what is noble in the first. They also appear to have something of the monkey, at least in the nose. Not made to suffer the bit in the mouth, the power of jaw is wanting. The determining marks concerning the bit are found between the eyes and the nose. No traces of courage or daring are found in these parts. The threatening snort of the ox and horse is not perceptible in these ape-like nostrils; none of the powers of plunder and prey in the feeble upper and under jaw. Nothing but burthen-bearing patience in the eyes.

Wild cruelty, the menacing power of rending, appear in the bear, abhorring man, the friend of ancient savage nature.

The most indolent, helpless, wretched creature, and of the most imperfect formation, is the *eunau ai*, or sloth. How extraordinary is the feebleness of the outline of the head, body, and feet! no sole of the feet, no toes small or great, which move independently, having but two or three long inbent claws, which can only move together. Its sluggishness, stupidity, and self-neglect are indescribable.

In the wild-boar every one may read ferocity, a want of all that is noble, greediness, stupidity, blunt feeling, gross appetite; and in the badger, ignoble, faithless, malignant, savage gluttony.

Remarkable is the profile of the lion, especially the outline of the forehead and nose. A man whose profile of forehead and nose should resemble that of the lion, would certainly be no common man; but such I have never seen. I own, the nose of the lion is much less prominent than that of man, but much more than that of any other quadruped. Royal, brutal strength, and arrogant usurpation, are evident; partly in the arching of the nose, partly in its breadth and parallel lines, and especially in the almost right angle, which the outline of the eyelid forms with the side of the nose.

In the eye and snout of the tiger, what bloodthirsty cruelty, what insidious craft! Can the laugh of Satan himself, at a fallen saint, be more fiend-like than the head of the triumphant tiger? Cats are tigers in miniature, with the advantage of domestic education. Little better in character, inferior in power. Unmerciful to birds and mice, as the tiger to the lamb. They delight in prolonging torture before they devour, and in this they exceed the tiger.

The more violent qualities of the elephant are discoverable in the number and size of his bones; his intelligence in the roundness of his form, and his docility in the massiness of his muscles; his art and discretion in the flexibility of his trunk; his retentive memory in

the size and arching of his forehead, which approaches
nearer to the outline of the human forehead than that of
any other beast. Yet how essentially different is it from
the human forehead, in the position of the eye and
mouth, since the latter generally makes nearly a right
angle with the axis of the eye and the middle line of
the mouth.

The crocodile proves how very physiognomical teeth
are. This, like other creatures, but more visibly and infallibly than others, in all its parts, outlines, and points,
has physiognomy that cannot be mistaken. Thus debased, thus despicable, thus knotty, obstinate, and wicked,
thus sunken below the noble horse, terrific, and void of
all love and affection, is this fiend incarnate.

Little acquainted as I am with horses, yet it seems to
me indubitable, that there is as great a difference in the
physiognomy of horses as in that of men. The horse
deserves to be particularly considered by the physiognomist, because it is one of those animals whose physiognomy, at least in profile, is so much more prominent,
sharp, and characteristic than that of most other beasts.

Of all animals the horse is that which, to largeness of
size, unites most proportion and elegance in the parts of
his body; for, comparing him to those which are immediately above or below him, we shall perceive that the
ass is ill made, the head of the lion is too large, the legs
of the ox too small, the camel is deformed, and the
rhinoceros and elephant too unwieldy. There is scarcely
any beast has so various, so generally marking, so speaking a countenance, as a beautiful horse.

"The upper part of the neck from which the mane
flows, in a well-made horse, ought to rise at first in a
right line; and, as it approaches the head, to form a

curve somewhat similar to the neck of the swan. The lower part of the neck ought to be rectilinear in its direction from the chest to the nether jaw, but a little inclined forward; for, were it perpendicular, the shape of the neck would be defective. The upper part of the neck should be thin and not fleshy; nor the mane, which ought to be tolerably full, and the hair long and straight. A fine neck ought to be long and elevated, yet proportionate to the size of the horse. If too long and small, the horse would strike the rider with his head; if too short and heavy, he would bear heavy on the hand. The head is advantageously placed when the forehead is perpendicular to the horizon. The head ought to be bony and small, not too long; the ears near each other, small, erect, firm, straight, free, and situated on the top of the head. The forehead should be narrow and somewhat convex, the hollows filled up; the eyelids thin; the eyes clear, penetrating, full of ardour, tolerably large as I may say, and projecting from the head, the pupil large, the under jaw bony, and rather thick; the nose somewhat arched, the nostrils open and well slit, the partition thin; the lips fine, the mouth tolerably large, the withers high and sharp." I must beg pardon for this quotation from the *Encyclopédie*, and for inserting thus much of the description of a beautiful horse, in a physiognomical essay intended to promote the knowledge and the love of man.

The more accurately we observe horses, the more shall we be convinced that a separate treatise of physiognomy might be written on them. I have somewhere heard a general remark, that horses are divided into three classes, the swan-necked, the stag-necked, and the hog-necked. Each of these classes has its peculiar countenance and character, and from the blending of which various others originate.

The heads of the swan-necked horses are commonly even, the forehead small, and almost flat; the nose extends, arching, from the eyes to the mouth; the nostrils are wide and open; the mouth small; the ears little, pointed, and projecting; the eyes large and round; the jaw below small; above, something broader; the whole body well proportioned, and the horse beautiful. This kind is cheerful, tractable, and high-spirited. They are very sensible of pain, which when dressing they sometimes express by the voice. Flattery greatly excites their joy, and they will express their pride of heart by parading and prancing. I will venture to assert that a man with a swan-neck, or, what is much more determinate, with a smooth projecting profile, and flaxen hair, would have similar sensibility and pride.

The stag-necked has something, in the make of his body, much resembling the stag itself. The neck is small, large, and scarcely bowed in the middle. He carries his head high. I have seen none of these. They are racers and hunters, being particularly adapted for swiftness by the make of the body.

The hog-necked. The neck above and below is alike broad; the head hanging downwards; the middle of the nose is concave in profile; the ears are long, thick, and hanging; the eyes small and ugly; the nostrils small, the mouth large, the whole body round, and the coat long and rough. These horses are intractable, slow, and vicious, and will run the rider against a wall, stone, or tree. When held in they rear, and endeavour to throw the rider. Blows or coaxing are frequently alike ineffectual; they continue obstinate and restive.

If we examine the different heads of horses, we shall find that all cheerful, high-spirited, capricious, courageous horses, have the nosebone of the profile convex; and that most of the vicious, restive, and idle, have the same bone flat or concave. In the eyes, mouth, and especially in the nostrils and jawbones, are remarkable varieties, concerning which I shall say nothing. I shall here add some remarks on the horse, communicated by a friend.

The grey is the tenderest of horses, and we may here add that people with light hair, if not effeminate, are yet, it is well known, of tender formation and constitution. The chestnut and iron grey, the black and bay, are hardy; the sorrel are the most hardy, and yet the most subject to disease. The sorrel, whether well or ill formed, is treacherous. All treacherous horses lay their ears on their neck. They stare and stop, and lay down their ears alternately.

The following passage, on the same subject, is cited from another writer: "When a horse has broad, long, widely separated, hanging ears, we are well assured he is bad and sluggish. If he lays down his ears alternately, he is fearful, and apt to start. Thin, pointed, and projecting ears, on the contrary, denote a horse of good disposition."

We never find that the thick, hog-necked horse is sufficiently tractable for the riding-house, or that he is of a strong nature when the tail shakes like the tail of a dog. We may be certain that a horse with large cheerful eyes, and a fine shining coat, if we have no other tokens, is of a good constitution and understanding.

These remarks are equally applicable to oxen and sheep, and probably to all other animals. The white ox is not so long serviceable for draught or labour as the black or red; he is more weak and silly than these. A sheep with short legs, strong neck, broad back, and cheerful eyes, is a good breeder, and remains peaceably with the flock. I am clearly of opinion that, if we may judge of the internal by the external of beasts, men may be judged of in the same manner.

CHAPTER XLI.

Of Birds, Fishes, Serpents, and Insects.

BIRDS.

BIRDS, whether compared to each other, or to other creatures, have their distinct characters. The structure of birds throughout is lighter than that of quadrupeds. Nature, ever steadfast to truth, thus manifests herself in the form of birds. Their necks are more pliant, their heads smaller, their mouths more pointed, and their garb more light and strong, than those of quadrupeds.

Their distinction of character, or gradation of passive and active power, is expressed by the following physiognomical varieties.

1. By the form of the skull. The more flat the skull the more weak, flexible, tender, and sensible is the character of the animal. This flatness contains less, and resists less.

2. By the length, breadth, and arching, or obliquity of their beaks. And here again we find, when there is arching, there is a greater extent of docility and capacity.

3. By the eyes, which appear to have an exact correspondence with the arching of the beak.

4. Particularly by the middle line, I cannot say of the mouth, but what is analogous to the mouth, the

beak; the obliquity of which is ever in a remarkable proportion with the outline of the profile of the head.

Who can behold the eagle hovering in the air, the powerful lord of so many creatures, without perceiving the seal, the native star of royalty, in his piercing round eye, the form of his head, his strong wings, his talons of brass, and in his whole form his victorious strength, his contemptuous arrogance, his fearful cruelty, and his ravenous propensity?

Consider the eyes of all living creatures, from the eagle to the mole; where else can be found that lightning glance which defies the rays of the sun?—where that capacity for the reception of light? How truly, how emphatically, to all who will hear and understand, is the majesty of his kingly character visible, not alone in his burning eye, but in the outline of what is analogous to the eyebone, and in the skin of the head, where anger and courage are seated? But throughout his whole form where are they not?

Compare the vulture with the eagle, and who does not observe in his lengthened neck and beak, and in his more extended form, less power and nobility than in the eagle? In the head of the owl, the ignoble greedy prey; in the.dove, mild, humble timidity; and in the swan, more nobility than in the goose, with less power than in the eagle, and tenderness than in the dove; more pliability than in the ostrich; and, in the wild-duck, a more savage animal than in the swan, without the force of the eagle?

Fish.

How different is the profile of a fish from that of a man?—how much the reverse of human perpendicu-

larity! How little is there of countenance when compared to the lion! How visible is the want of mind,
reflection, and cunning! What little or no analogy to
forehead! What an impossibility of covering or entirely
closing the eyes! The eye itself is merely circular and
prominent, has nothing of the lengthened form of the
eye of the fox or elephant.

Serpents.

I will allow physiognomy, when applied to man, to
be a false science, if any being throughout nature can
be discovered void of physiognomy, or a countenance
which does not express its character. What has less,
yet more, physiognomy than the serpent? May we not
perceive in it tokens of cunning and treachery? Certainly not a trace of understanding or deliberate plan.
No memory, no comprehension, but the most unbounded
craft and falsehood. How are these reprobate qualities
distinguished in their forms? The very play of their
colours, and wonderful meandering of their spots, appear
to announce and to warn us of their deceit.

All men possessed of real power are upright and
honest; craft is but the substitute of power. I do not
here speak of the power contained in the folds of a
serpent; they all want the power to act immediately,
without the aid of cunning. They are formed to " bruise
the heel, and to have the head bruised." The judgment
which God has pronounced against them is written on
their flat, impotent forehead, mouth, and eyes.

Insects.

How inexpressibly various are the characteristics impressed by the eternal Creator on all living beings!

How has he stamped on each its legible and peculiar properties! How especially visible is this in the lowest classes of animal life! The world of insects is a world of itself. The distance between this and the world ot men, I own, is great; yet were it sufficiently known, how useful would it be to human physiognomy! What certain proofs of the physiognomy of men must be obtained from insect physiognomy!

How visible are their powers of destruction, of suffering and resisting, of sensibility and insensibility, through all their forms and gradations! Are not all the compact, hard-winged insects physiognomically and characteristically more capable and retentive than the various light and tender species of the butterfly? Is not the softest flesh the weakest, the most suffering, the easiest to destroy? Are not the insects of least brains the beings most removed from man, who has the most brain? Is it not perceptible in each species whether it be warlike, defensive, enduring, weak, enjoying, destructive, easy to be crushed, or crushing? How distinct in the external character are their degrees of strength, of defence, of stinging, or of appetite!

The great dragon fly shows its agility and swiftness, in the structure of its wings; perpetually on flight in search of small flies. How sluggish, on the contrary, is the crawling caterpillar! How carefully does he set his feet as he ascends a leaf! How yielding his substance, incapable of resistance! How peaceable, harmless, and indolent is the moth! How full of motion, bravery, and hardiness is the industrious ant! How loath to remove on the contrary, is the harnessed lady-bird!

CHAPTER XLII.

On Shades.

THOUGH shades are the weakest and most vapid, yet they are at the same time, when the light is at a proper distance, and falls properly on the countenance, to take the profile accurately, the truest representation that can be given of man. The weakest, for it is not positive; it is only something negative, only the boundary line of half the countenance. The truest, because it is the immediate expression of nature, such as not the ablest painter is capable of drawing by hand after nature. What can be less the image of a living man than a shade? Yet how full of speech? Little gold, but the purest.

The shade contains but one line; no motion, light, colour, height, or depth; no eye, ear, nostril, or cheek; but a very small part of the lip; yet how decisively it is significant! Drawing and painting, it is probable, originated in shades. They express, as I have said, but little; but the little they do express is exact. No art can attain to the truth of the shade taken with precision. Let a shade be taken after nature with the greatest accuracy, and with equal accuracy be afterwards reduced upon fine transparent oil-paper. Let a profile of the same size be taken by the greatest master in his happiest moment, then let the two be laid upon each other, and the difference will be immediately evident.

I never found, after repeated experiments, that the best efforts of art could equal nature either in freedom or in precision, but that there was always something more or less than nature. Nature is sharp and free; whoever studies sharpness more than freedom, will be hard, and whoever studies freedom more than sharpness will become diffuse and indeterminate. I can admire him only who, equally studious of her sharpness and freedom, acquires equal certainty and impartiality.

To attain this, artist, imitator of humanity! first exercise yourself in drawing shades; afterwards copy them by hand, and next compare and correct. Without this you will with difficulty discover the grand secret of uniting precision and freedom.

I have collected more physiognomical knowledge from shades alone than from every other kind of portrait ; have improved physiognomical sensation more by the sight of them than by the contemplation of ever mutable nature. Shades collect the distracted attention, confine it to an outline, and thus render the observation more simple, easy, and precise. Physiognomy has no greater, more incontrovertible certainty of the truth of its object, than that imparted by shade. If the shade, according to the general sense and decision of all men, can decide so much concerning character, how much more must the living body, the whole appearance, and action of the man ! If the shade be oracular, the voice of truth, the word of God, what must the living original be, illuminated by the spirit of God !

Hundreds have asked, and hundreds will continue to ask, " What can be expected from mere shades ?" Yet no shade can be viewed by any one of these hundred, who will not form some judgment on it, often accurately, more accurately than I could have judged.

In order to make the astonishing significance of shades conspicuous, we ought either to compare opposite characters of men taken in shade, or, which may be more convincing, to cut out of black paper, or draw, imaginary countenances widely dissimilar. Or, again, when we have acquired some proficiency in observation, to double black paper, and cut two countenances; and afterwards, by cutting with the scissors, to make slight alterations, appealing to our eye, or physiognomical feeling, at each alteration; or, lastly, only to take various shades of the same countenance, and compare them together. Such experiments would astonish us, to perceive what great effects are produced by slight alterations.

The common method of taking shades is accompanied with many inconveniences. It is hardly possible the person drawn should sit sufficiently still; the designer is obliged to change his place; he must approach so near to the person that motion is almost inevitable, and the designer is in the most inconvenient position; neither are the preparatory steps every where possible, nor simple enough. A seat purposely contrived would be more convenient. The shade should be taken on post paper, or rather on thin oil-paper, well dried. Let the head and back be supported by a chair, and the shade fall on the oil-paper behind a clear, flat, polished glass. Let the drawer sit behind the glass, holding the frame with his left hand, and, having a sharp black-lead pencil, draw with the right. The glass, in a detached sliding frame, may be raised or lowered, according to the height of the person. The bottom of the glass frame, being thin, will be best of iron, and should be raised so as to rest steadily upon the shoulder. In the centre, upon the glass, should be a small piece of wood or iron, to which fasten a small round cushion, supported by a short pin, scarcely half an inch long, which also may be raised or lowered, and against which the person drawn may lean.

CHAPTER XLIII.

Description of Plate VI.

Number I. MENDELSSOHN.

IN the forehead and nose, penetration and sound understanding are evident.   The mouth is much more delicate than the mouth of 2.

Number II. J. SPALDING.

Clear ideas, love of elegance, purity, accuracy of thought and action ; does not easily admit the unnatural. The forehead not sufficiently characteristic, but fine taste in the nose.

Number III. ROCHOW.

Has more good sense ; prompt, accurate perception of truth and delicacy, than 4; but I suspect less acuteness.

Number IV. P. NICOLAI.

Whoever hesitates concerning the character of this head, never can have observed the forehead.   This arch, abstractedly considered, especially in the upper part, has more capacity than Nos. 2 and 3.   In the upper outline, also, of the under part, understanding and exquisite penetration cannot be overlooked.

Number V.

One of those masculine profiles which generally please. Conceal the under chin, and an approach to greatness is perceptible ; except that greater variation in the outline is wanting, especially in the nose and forehead. The choleric phlegmatic man is visible in the whole ; especially in the eyebrows, nose, and bottom part of the

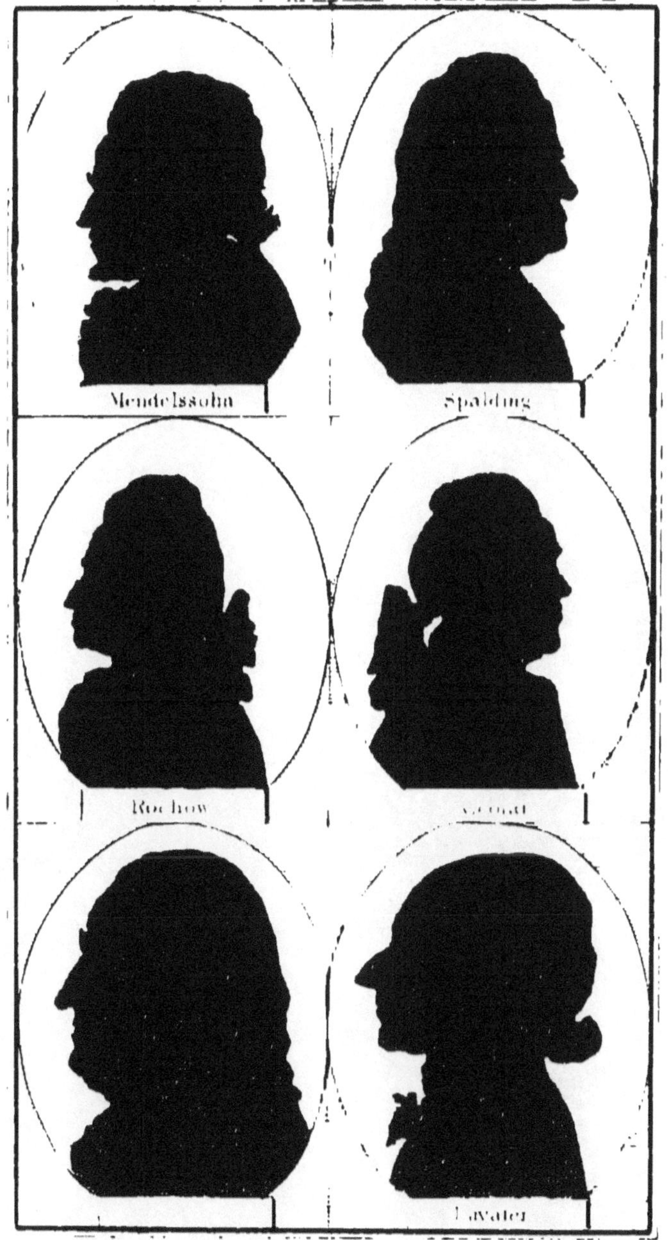

Mendelssohn
Spalding
Rochow
Gedal
Lavater

chin; as likewise are integrity, fidelity, goodness, and complaisance.

Number VI. J. C. LAVATER.

This shade, though imperfect, may easily be known. It must pass without comment, or rather the commentary is before the world—is in this book. Let that speak; I am silent.

CHAPTER XLIV.

A Word to Travellers.

THERE appear to me to be three things indispensable to travellers—health, money, and physiognomy. Therefore a physiognomical word to travellers. I could wish indeed, that, instead of a word, *a traveller's physiognomical companion* were written; but this must be done by an experienced traveller. In the mean time I shall bid him farewell, with the following short advice :—

What do you seek, travellers? what is your wish? What would you see more remarkable, more singular, more rare, more worthy to be examined, than the varieties of humanity? This indeed is fashionable. You inquire after men; you seek the wisest, best, and greatest men, especially the most famous. Why is your curiosity limited to seeing only? Would it not be better you should illuminate your own minds by the light of others, and animate yourselves by their ardour?

His curiosity is childish which is merely confined to seeing, whose ambition desires only to say, I have beheld that man. He who would disregard views so confined, must study such men physiognomically; if he would learn wisdom, he must be able to compare and judge of the relation between their works, their fame, and their form. By this only may much be learned. By this may the stream be compared to the fountain, the quality of the waters examined, their course, their gentle murmurs, or more boisterous war. The inquirer may ask, what is the degree of originality of those men, what is borrowed, what is internal, what external? This forehead and these eyebrows will thus versify, thus translate, thus criticise; therefore on this eye depends the fate of the writer, the blockhead, or the man of genius. This nose thus estimates the mortal and the immortal in the human performances. As are the features so will be the mind.

Yes, scholars of nature, you have much to learn from the countenances of famous men. In them you will read that the wasp will dare to alight on the nose of the hero. To me it will be pleasure when you have acquired this physiognomical sensation; for, without this, you will but travel in the dark; you will but be led through a picture-gallery blindfold, only that you might say, I too have been in that gallery.

Could I travel unknown, I would also visit artists, men of learning, and philosophers, men famous in their respective countries; but it should either be my adieu, as the thing least important, or as a recreation on my arrival. Pardon me, men of renown; I have been credulous in your favour, but I daily become more circumspect. Far be it from me to depreciate your worth. I know many whose presence does not diminish but increase fame; yet will I be careful that remorse shall neither dazzle nor cloud my reason.

It would be much more agreeable to me to mix unknown with the multitude, visit churches, public walks, hospitals, orphan-houses, and assemblies of ecclesiastics and men of the law. I would first consider the general form of the inhabitants, their height, proportion, strength, weakness, motion, complexion, attitude, gesture, and gait. I would observe them individually, see, compare, close my eyes, trace in imagination all I have seen, open them again, correct my memory, and close and open them alternately. I would study for words, write, and draw, with a few determinate traits, the general form, so easy to be discovered. I would compare my drawings with the known general form of the people. How easily might a summary, an index of the people, be obtained!

Having made these familiar to me, I would descend to the particular, would search for the general form of the head, would ask, Is it most confined to the cylindrical, the spherical, the square, the convex, or the concave? Is the countenance open, is it writhed, is it free, or forked? I would next examine the forehead, then the eyebrows, the outline and colour of the eyes, the nose, and especially the mouth when it is open; and the teeth, with their appearances, to discover the national characteristic.

Could I but define the line of the opening of the lips in seven promiscuous countenances, I imagine I should have found the general physiognomical character of the nation or place. I almost dare to establish it as an axiom, that what is common to six or seven persons of any place, taken promiscuously, is more or less common to the whole. Exceptions there may be, but they will be rare.

In the next place, I would plant myself in a public walk, or at the crossing of streets. There I would wait patiently for the unknown noble countenance, uncorrupted by fame and adulation, which certainly, most certainly, I should find: for, in all countries on earth wherever a hundred common men are assembled, one not common may be found; and out of a thousand, ten.

I must have indeed little eye, little sensibility for noble humanity, little faith in Providence, which seeks its adorers, if I did not find this one in a hundred, or at least in the ten among a thousand. He that seeketh shall find. I waited not in vain. He came, I found him, he passed by me. And what were the tokens by which I discovered him in every town, in every nation, under every cope of heaven, and among all people, kindred, and tongues? By the general combination of the countenance, by the upper outline of the forehead, the eyebrows, the basis of the nose, and the mouth, so conformable to each other, so parallel and horizontal, at the first glance. By the wrinkless, compressed yet open forehead, the powerful eyebrows; the easily discerned, easily delineated space between the eyebrows, which extends itself to the back of the nose, like the great street from the market-place to the chief gate of a city. By the shut but freely breathing mouth; the chin neither haggard nor fleshy; the deep and shining attraction of the eye; which all, incautiously and unintentionally, betrayed themselves to my research; or, I discovered him even in his foreign and distorted form, from which the arrogant, self-supposed handsome, would turn with contempt. I see through his disguise, as I should the hand of a great master through the smear of varnish.

I approach the favourite of heaven. I question him concerning what I do, and what I do not wish to know, that I may hear the voice of the soul proceeding from

the mouth ; and, viewing him nearer, I see all the obli-
quities of distortion vanish.   I ask him concerning his
occupation, his family, his place of residence.   I inquire
the road thither.   I come unexpectedly upon him into
his house, into his workshop; he rises, I oblige him to
be seated, to continue his labour.   I see his children, his
wife, and am delighted.   He knows not what I want,
nor do I know myself, yet am I pleased with him, and
he with me.   I purchase something or nothing, as it
happens.   I inquire particularly after his friends.   "You
have but few, but those few are faithful."   He stands
astonished, smiles or weeps, in the innocence and goodness of his heart, which he wishes to conceal, but which
is open as day.   He gains my affection ; our emotions
are reciprocally expanded and strengthened ; we separate
reluctantly, and I know I have entered a house which is
entered by the angels of God.

Oh ! how gratefully, how highly is he rewarded for
his labours who travels, interested in behalf of humanity,
and, with the eyes of a man, to collect in the spirit
the children of God, who are scattered over the world !
This appears to me to be the supreme bliss of man, as it
must be of angels.

If I do not meet him, I have no resource but in
society.   Here I hear him most who speaks least, mildest, and most unaffectedly.   Wherever I meet the smile
of self-sufficiency, or the oblique look of envy, I turn
away, and seek him who remains oppressed by the loud
voice of confidence.   I set myself rather beside the
answerer than the man of clamorous loquacity; and still
rather beside the humble inquirer than the voluble
solver of all difficulties.

He who hastens too fast, or lags behind, is no companion of mine. I rather seek him who walks with a free, firm, and even step; who looks but little about him; who neither carries his head aloft, nor contemplates his legs and feet. If the hand of affliction be heavy on him, I set myself by his side, take his hand, and, with a glance, infuse conviction to his soul, that God is love.

In my memory I retain the simple outlines of the loud and the violent, the laughter and the smiles, of him who gives the key, and him who takes. I then commit them to paper; my collection increases. I compare, arrange, judge, and am astonished. I every where find similarity of traits, similarity of character; the same humanity every where, and every where the same tokens.

CHAPTER XLV.

A Word to Princes and Judges.

FOR your use, most important of men, how willingly would I write a treatise! Who, so much as you, need a perfect knowledge of man, free from cabal or the intervention of self-interest? Suffer me to approach your throne, and present my address.

In your most secret commonplace-book keep an index to each class of character among men, taken from at least ten of the most accurate proofs; not at a distance, not among foreigners, but seek at home for the wisest and best of your own subjects. Wherever a wise and good prince governs, there are excellent subjects. Such a prince believes that he has such subjects, although at the moment he should be unacquainted with them; or, at least, that he has subjects capable of wisdom and goodness. Wherever one good person is, there certainly are two, as certainly as where the female is there will the male be.

Suffer me, princes, consecrated as you are among men, to entreat you, for the honour of humanity, principally to study, to seek for, and to seize on excellence. Judge not too suddenly, nor by mere appearances. That which a prince once approves, it may afterwards be difficult or dangerous to reject. Depend not on the testimony of others, which, to princes especially, is ever exaggerated either in praise or blame; but examine the countenance, which, though it may dissemble to a prince, or rather to the dignity of a prince, cannot deceive him as a man. Having once discovered wisdom and goodness in a subject, honour such a subject as the best blessing which Heaven can in this world bestow upon its favourites. Seek features that are strong, but not forbidding; gentle, yet not effeminate; positive, without turbulence; natural, not arrogant; with open eyes, clear aspects; strong noses near the forehead, and with such let your thrones be surrounded.

Intrust your secrets to proportionate and parallel drawn countenances; to horizontal, firm, compressed eyebrows; channelled, not too rigorously closed, red, active, but not relaxed or withered lips. Yet I will forbear to delineate, and again only entreat that the countenance may be sacred to you for the sake of goodness and wisdom.

As to you, judges, judge not indeed by appearances, but examine according to appearances. Justice blindfold without physiognomy, is as unnatural as blindfold love. There are countenances which cannot have committed a multitude of vices. Study the traits of each vice, and the forms in which vice naturally or unwillingly resides. There are capabilities and incapabilities in the countenance, things which it can will, others which it cannot. Each passion, open or concealed, has its peculiar language. The appearance of innocence is as determinate to the experienced eye as the appearance of health.

Bring guilt and innocence face to face, and examine them in your presence, and when they suppose you do not observe them in the presence and in the absence of witnesses ; with justice see, with justice hear and obey, the determined voice of unprejudiced conviction. Remark their walk when they enter, and when they leave the judgment-hall. Let the light fall upon their countenances ; be yourself in the shade. Physiognomy will render the torture unnecessary,* will deliver innocence, will make the most obdurate vice turn pale, will teach us how we may act upon the most hardened. Every thing human must be imperfect; yet will it be evident that the torture, more disgraceful to man than the halter, the axe, and the wheel, is infinitely more uncertain and dangerous than physiognomy. The pain of torture is more horrible even than the succeeding death, yet it is only to prove, to discover truth. Physiognomy shall not execute, and yet it shall prove ; and by its proof vice alone, and not innocence, shall suffer. O ye judges of men, be men, and humanity shall teach you, with more open eyes, to see and abhor all that is inhuman !

* A few years since one philosopher wrote to another, The *torture* will soon be abolished in Austria. It was asked, What shall be its substitute? The penetrating look of the judge, replied Sonnenfels. Physiognomy will, in twenty-five years, become a part of jurisprudence instead of torture, and lectures will be read in the universities on the *Physiognomice forense* instead of the *Medicina forensis.*

CHAPTER XLVI.

A Word to the Clergy.

You also, my brethren, need a certain degree of physiognomy, and perhaps, princes excepted, no men more. You ought to know whom you have before you, that you may discern spirits, and portion out the word of truth to each, according to his need and capacity. To whom can a knowledge of the degree of actual and possible virtue, in all who appear before you, be more advantageous than to you?

To me physiognomy is more indispensable than the liturgy. It is to me alike profitable for doctrine, exhortation, comfort, correction, examination; with the healthy, with the sick, the dying, the malefactor; in judicial examinations, and the education of youth. Without it, I should be as the blind leading the blind.

I might be robbed of my ardour or inspired with enthusiasm by a single countenance. Whenever I preach, I generally seek the most noble countenance, on which I endeavour to act, and the weakest when teaching children. It is generally our own fault if our hearers are inattentive, if they do not themselves give the key in which it is necessary they should be addressed.

Every teacher possessed of physiognomical sensation will easily discern and arrange the principal classes among his hearers, and what each class can and cannot receive. Let six or seven classes, of various capacities, be selected; let a chief, a representative, a characteristic countenance of each class be chosen : let these countenances be fixed in the memory, and let the preacher accommodate himself to each ; speaking thus to one, and thus to another, and in such a manner to a third.

R

There cannot be a more natural, effective, or definite incitement to eloquence, than supposing some characteristic countenance present, of the capacity of which almost mathematical certainty may be obtained. Having six or seven, I have nearly my whole audience before me. I do not then speak to the winds. God teaches us by physiognomy to act upon the best of men according to the best of means.

CHAPTER XLVII.

Physiognomical Elucidations of Countenances.

A regular well-formed countenance is where all the parts are remarkable for their symmetry; the principal features, as the eyes, nóse, and mouth, neither small nor bloated. In which the position of the parts, taken together and viewed at a distance, appears nearly horizontal and parallel.

A beautiful countenance is that in which, besides the proportion and position of the parts, harmony, uniformity, and mind are visible; in which nothing is superfluous, nothing deficient, nothing disproportionate, nothing super-added, but all is conformity and concord.

A pleasant countenance does not necessarily require perfect symmetry and harmony; yet nothing must be wanting, nothing burdensome. Its pleasantry will principally exist in the eye and lips, which must have nothing commanding, arrogant, contemptuous, but must generally speak complacency, affability, and benevolence.

A gracious countenance arises out of the pleasant, when, far from any thing assuming, to the mildest benevolence are added affability and purity.

A charming countenance must not simply consist

either of the beautiful, the pleasant, or the gracious ; but when to these is added a rapid propriety of motion, which renders it charming.

An insinuating countenance leaves no power to active or passive suspicion. It has something more than the pleasant, by infusing that into the heart which the pleasant only manifests.

Other species of these delightful countenances are, the attracting, the winning, the irresistible.

Very distinct from all these are the amusing, the divertingly loquacious, the merely mild, and also the tender and delicate.

Superior, and more lovely still, is the purely innocent, where no distorted oblique muscle, whether in motion or at rest, is ever seen.

This is still more exalted when it is full of soul, of natural symapthy, and power to excite sympathy.

When in a pure countenance good power is accompanied by a spirit of order, I may call it an Attic countenance.

Spiritually beautiful may be said of a countenance where nothing thoughtless, inconsiderate, rude, or severe, is to be expected; and the aspect of which immediately and mildly incites emotion in the principal powers of the mind.

Noble is when we have not the least indiscretion to fear, and when the countenance is exalted above us, without a possibility of envy; while it is less sensible of its own superiority than of the pleasure we receive in its presence.

A great countenance will have few small secondary traits; will be in grand divisions, without wrinkles; must exalt, must affect us, in sleep, in plaster of Paris,

in every kind of caricatures; as, for example, that of
Philip de Comines.

A sublime countenance can neither be painted nor described; that by which it is distinguished from all others
can only be felt. It must not only move, it must exalt
the spectator. We must at once feel ourselves greater
and less in its presence than in the presence of all others.

Whoever is conscious of its excellence, and can
despise or offend it, may, as hath been before said,
blaspheme against the great Author of his existence.

CHAPTER XLVIII.

Physiognomical Anecdotes.

1.

I HAVE nothing to require of you, said a father to his
innocent son, when bidding him farewell, but that you
bring me back your present countenance.

2.

A noble, amiable, and innocent young lady, who had
been educated principally in the country, saw her face
in the glass as she passed it with a candle in her hand,
retiring from evening prayers, and having just laid
down her Bible. Her eyes were cast to the ground with
inexpressible modesty at the sight of her own image.
She passed the winter in town, surrounded by adorers,
hurried away by dissipation, and plunged in trifling
amusements. She forgot her Bible and her devotion.
In the beginning of spring she returned to her country
seat, her chamber, and the table on which the Bible lay.
Again she had the candle in her hand, and again saw
herself in the glass. She turned pale, put down the

candle, retreated to a sofa, and fell on her knees: "O God! I no longer know my own face. How am I degraded! My follies and vanities are all written in my countenance. Wherefore have they been neglected, illegible, to this instant? O come and expel, come and utterly efface them, mild tranquillity, sweet devotion, and ye gentle cares of benevolent love!"

3.

"I will forfeit my life," said Titus of the priest Tacitus, "if this man be not an arch knave. I have three times observed him sigh and weep without cause; and ten times turn aside to conceal a laugh he could not restrain, when vice or misfortune were mentioned."

4.

A stranger said to a physiognomist, "How many dollars is my face worth?"—"It is hard to determine," replied the latter.—"It is worth fifteen hundred," continued the questioner, "for so many has a person lent me upon it, to whom I was a total stranger."

5.

A poor man asked alms, "How much do you want?" said the person of whom he asked, astonished at the peculiar honesty of his countenance. "How shall I dare to fix a sum?" answered the needy person. "Give me what you please, sir, I shall be contented and thankful."—"Not so," replied the physiognomist; "as God lives, I will give you what you want, be it little or much." "Then, sir, be pleased to give me eight shillings."—"Here they are; had you asked a hundred guineas you should have had them."

CHAPTER XLIX.

Miscellaneous Extracts from Kœmpf's Essay on the Temperaments, with Remarks.

1.

"WILL not physiognomy be to man what the looking-glass is to an ugly woman?"

Let me also add, to the handsome woman. The wise looks in the glass, and washes away spots; the fool looks, turns back, and remains as he was.

2.

"Each temperament, each character, has its good and bad. The one has inclinations of which the other is incapable. The one has more than the other. The ingot is of more worth than the guineas individually into which it is coined; yet the latter are most useful. The tulip delights by its beauty, the carnation by its smell. The unseemly wormwood displeases both taste and smell, yet in medical virtue is superior to both. There it is that each contributes to the perfection of the whole."

The carnation should not wish to be a tulip, the finger an eye, nor the weak desire to act within the circle of the strong. Each has its peculiar circle, as it has its peculiar form. To wish to depart from this circle is like wishing to be transported into another body.

3.

"Within the course of a year we are assured that the activity of nature changes the body, yet we are sensible of no change of mind, although our body has been subjected to the greatest changes, in consequence of meat, drink, air, and other accidents; the difference of air and manner of life does not change the temperament."

The foundation of character lies deeper, and is, in a certain degree, independent of all accidents. It is probably the spiritual and immortal texture into which all that is visible, corruptible, and transitory is interwoven.

4.

"A block of wood may be carved by a statuary into what form he shall please; he may make it an Æsop or an Antinous, but he will never change the inherent nature of the wood."

To know and distinguish the materials and form of men, so far as knowledge contributes to their proper application, is the highest and most effectual wisdom of which human nature is capable.

5.

"In the eyes of certain persons there is something sublime, which beams and exacts reverence. This sublimity is the concealed power of raising themselves above others, which is not the wretched effect of constraint, but primitive essence. Each finds himself obliged to submit to this secret power without knowing why, as soon as he perceives that look, implanted by nature to inspire reverence, shining in the eyes. Those who possess this natural, sovereign essence, rule as lords or lions among men by native privilege, with heart and tongue conquering all."

6.

"There are only four different aspects, all different from each other, the ardent, the dull, the fixed, and the fluctuating."

The application is the proof of all general propositions. Let physiognomical axioms be applied to known individuals, friends or enemies, and their truth or falsehood, precision or inaccuracy, will easily be determined. Let us make the experiment with the above, and we shall certainly find there are numerous aspects which are not included within these four; such as the luminous aspect, very different from the ardent, and neither fixed like the melancholic, nor fluctuating like the sanguine.

There is the look or aspect which is at once rapid and fixed, and, as I may say, penetrates and attaches at the same moment. There is the tranquilly active look, neither choleric nor phlegmatic. I think it would be better to arrange them into the giving, receiving, and the giving and receiving combined; or, into intensive and extensive; or, into the attracting, repelling, and unparticipating; into the contracted, the relaxed, the strained, the attaining, the unattaining, the tranquil, the steady, the slow, the open, the closed, the cold, the amorous, &c.

CHAPTER L.

Upon Portrait Painting.

PORTRAIT painting, the most natural, manly, useful, noble, and, however apparently easy, is the most difficult of the arts. Love first discovered this heavenly art. Without love, what could it perform?

As on this art depends a great part of this present work, and the science on which it treats, it is proper that something should be said on the subject. Something; for how new, how important, and great a work might be written on this art! For the honour of man, and of the art, I hope such a work will be written. I do not think it ought to be the work of a painter, however great in his profession, but of the understanding friend of physiognomy, the man of taste, the daily confidential observer of the great portrait painter.

Sultzer, that philosopher of taste and discernment, has an excellent article in his dictionary on this subject, under the word *Portrait.* But what can be said, in a work so confined, on a subject so extensive? Again, whoever will employ his thoughts on this art, will find that it is sufficient to exercise all the searching, all the active powers of man; that it never can be entirely learned, nor ever can arrive at ideal perfection.

I shall now attempt to recapitulate some of the avoidable and unavoidable difficulties attendant on this art; the knowledge of which, in my opinion, is as necessary to the painter as to the physiognomist.

Let us first inquire, What is portrait painting? It is the communication, the preservation of the image of some individual; the art of suddenly depicting all that can be depicted of that half of man which is rendered apparent, and which never can be conveyed in words. If what Goethe has somewhere said be true, and in my opinion nothing can be more true, that the best text for a commentary on man is his presence, his countenance, his form; how important, then, is the art of portrait painting!

To this observation of Goethe's, I will add a passage on the subject from Sultzer's excellent dictionary: "Since no object of knowledge whatever can be more important to us than a thinking and feeling soul, it cannot be denied but that man, considered according to his form, even though we should neglect what is wonderful in him, is the most important of visible objects."

The portrait painter should know, feel, and be penetrated with this: penetrated with reverence for the

greatest works of the greatest masters. Were such the
subject of his meditation, not from constraint, but native
sensation; were it as natural to him as the love of
life, how important, how sacred to him, would this art
become! Sacred to him should be the living countenance as the text of holy scripture to the translator. As
careful should the one be not to falsify the work, as
should be the other not to falsify the word of God.

Great is the contempt which an excellent translator
of an excellent work deserves, whose mind is wholly
inferior to the mind of his original. And is it not the
same with the portrait painter? The countenance is
the theatre on which the soul exhibits itself: here must
its emanations be studied and caught. Whoever cannot
seize these emanations cannot paint; and whoever cannot paint these is no portrait painter.

Each perfect portrait is an important painting, since
it displays the human mind with the peculiarities of
personal character. In such we contemplate a being
where understanding, inclinations, sensations, passions,
good and bad qualities of mind and heart, are mingled
in a manner peculiar to itself. We here frequently see
them better than in nature herself, since in nature nothing is fixed, all is swift, all is transient. In nature,
also, we seldom behold the features under that propitious aspect in which they will be transmitted by the
able painter.

If we could, indeed, seize the fleeting transitions of
nature, or had she her moments of stability, it would then
be much more advantageous to contemplate nature than
her likeness; but this being impossible, and since, likewise, few people will suffer themselves to be observed
sufficiently to deserve the name of observation, it is to

me indisputable, that a better knowledge of man may be obtained from portraits than from nature, she being thus uncertain, thus fugitive.

The rank of the portrait painter may hence be easily determined; he stands next to the historical painter Nay, history painting itself derives a part of its value from its portraits; for expression, one of the most important requisites in historical painting, will be the more estimable, natural, and strong, the more of natural physiognomy is expressed in the countenances, and copied after nature. A collection of excellent portraits is highly advantageous to the historical painter for the study of expression.

Where shall we find the historical painter who can represent real beings with all the decorations of fiction? Do we not see them all copying copies? True it is, they frequently copy from imagination; but this imagination is only stored with the fashionable figures of their own or former times.

Having presumed thus far, let us now enumerate some of the surmountable difficulties of portrait painting. I am conscious the freedom with which I shall speak my thoughts will offend, yet to give offence is far from my intention. I wish to aid, to teach that art, which is the imitation of the works of God: I wish improvement. And how is improvement possible without a frank and undisguised discovery of defects!

In all the works of portrait painters which I have seen, I have remarked the want of a more philosophical, that is to say, a more just, intelligible, and universal knowledge of men. The insect painter, who has no accurate knowledge of insects, the form, the general, the particular, which is appropriated to each insect, however

good a copyist he may be, will certainly be a bad painter
of insects. The portrait painter, however excellent a
copyist, (a thing much less general than is imagined by
connoisseurs,) will paint portraits ill if he have not the
most accurate knowledge of the form, proportion, connection, and dependence of the great, and minute parts of
the human body, as far as they have a remarkable influence on the superficies; if he has not most accurately
investigated each individual member and feature. For
my own part, be my knowledge what it may, it is far
from accurate in what relates to the minute specific
traits of each sensation, each member, each feature; yet
I daily remark that this acute, this indispensable knowledge, is at present every where uncultivated, unknown,
and difficult to convey to the most intelligent painters.

Those who will be at the trouble of considering a
number of men promiscuously taken, feature by feature,
will find that each ear, each mouth, notwithstanding
their infinite diversity, have yet their small curves,
corners, characters, which are common to all, and which
are found stronger or weaker, more or less marking, in all
men who are not monsters born, at least in these parts.

Of what advantage is all our knowledge of the great
proportions of the body and countenance? (Yet even
that part of knowledge is, by far, not sufficiently studied,
not sufficiently accurate. Some future physiognomical
painter will justify this assertion, till when, be it considered as nothing more than cavil.) Of what advantage, I say, is all our knowledge of the great proportions,
when the knowledge of the finer traits, which are equally
true, general, determinate, and no less significant, is
wanting? And this want is so great, that I appeal to
those who are best informed, whether many of the ablest

painters, who have painted numerous portraits, have any tolerably accurate or general theory of the mouth only. I do not mean the anatomical mouth, but the mouth of the painter, which he ought to see, and may see, without any anatomical knowledge.

I have examined volume after volume of engravings of portraits after the greatest masters, and am therefore entitled to speak. But let us confine observations to the mouth. Having previously studied infants, boys, youth, manhood, old age, maidens, wives, matrons, with respect to the general properties of the mouth; and having discovered these, let us compare, and we shall find that almost all painters have failed in the general theory of the mouth; that it seldom happens, and seems only to happen by accident, that any master has understood these general properties. Yet how indescribably much depends on them! What is the particular, what the characteristic, but shades of the general! As it is with the mouth, so it is with the eyes, eyebrows, nose, and each part of the countenance.

The same proportion exists between the great features of the face; and as there is this general proportion in all countenances, however various, so is there a similar proportion between the small traits of these parts. Infinitely varied are the great features in their general combination and proportion. As infinitely varied are the shades of the small traits in these features, however great their general resemblance. Without an accurate knowledge of the proportion of the principal features, as, for example, of the eyes and mouth, to each other, it must ever be mere accident, an accident that indeed rarely happens, when such proportion exists in the works of the painter Without an accurate knowledge of the particular constituent parts and traits of each principal feature, I once again repeat, it must be accident, miraculous accident, should any one of them be justly delineated.

The reflecting artist may be induced from this remark to study nature intimately by principle, and to show him, if he be in search of permanent fame, that, though he ought to behold and study the works of the greatest masters with esteem and reverence, he yet ought to examine and judge for himself. Let him not make the virtue modesty his plea, for under this does omnipresent mediocrity shelter itself. Modesty, indeed, is not so properly virtue as the garb and ornament of virtue, and of existing positive power. Let him, I say, examine for himself, and study nature in whole and in part, as if no man ever had observed, or ever should observe, but himself. Deprived of this, young artist, thy glory will but resemble a meteor's blaze; it will only be founded on the ignorance of your contemporaries.

By far the greater part of the best portrait painters, when most successful, like the majority of physiognomists, content themselves with expressing the character of the passions in the moveable, the muscular features of the face. They do not understand, they laugh at, rules which prescribe the grand outline of the countenance as indispensable to portrait painting, independent of the effects produced by the action of the muscles.

Till institutions shall be formed for the improvement of portrait painting, perhaps till a physiognomical society or academy shall produce physiognomical portrait painters, we shall at best but creep in the regions of physiognomy, where we might otherwise soar. One of the greatest obstacles to physiognomy is the actual,

incredible imperfection of this art. There is generally a
defect of eye or hand of the painter, or the object is
defective which is to be delineated, or perhaps all three.
The artist cannot discover what *is*, or cannot draw it
when he discovers it. The object continually alters its
position, which ought to be so exact, so continually the
same ; or should it not, and should the painter be endowed with an all-observing eye, an all-imitative hand,
still there is the last insuperable difficulty, that of the
position of the body, which can but be momentary,
which is constrained, false, and unnatural, when more
than momentary.

Trifling, indeed, is what I have said to what might be
said. According to the knowledge I have of it, this is
yet uncultivated ground. How little has Sultzer himself said on the subject! But what could he say in a
dictionary ? A work wholly dedicated to this is necessary to examine and decide on the works of the best
portrait painters, and to insert all the cautions and
rules necessary for the young artist, in consequence of
the infinite variety, yet incredible uniformity, of the
human countenance.

The artist who wishes to paint portraits perfectly,
must so paint that each spectator may with truth exclaim,
"This is indeed to paint! this is true, living likeness ;
perfect nature ; it is not painting! Outline, form, proportion, position, attitude, complexion, light and shade,
freedom, ease, nature ! Nature in every characteristic
disposition ! Nature in the complexion, in each trait, in
her most beauteous, happiest moments, her most select,
most propitious state of mind ; near at a distance, on
every side Truth and Nature ! Evident to all men, all
ages, the ignorant and the connoisseur; most conspicuous

to him who has most knowledge; no suspicion of art;
a countenance in a mirror, to which we would speak; that
speaks to us; that contemplates more than it is contem-
plated; we rush to it, we embrace it, we are enchanted !"

Young artist! emulate such excellence, and the least
of your attainments in this age will be riches and honour,
and fame in futurity. With tears you will receive the
thanks of father, friend, and husband, and your work
will honour that Being whose creation is the noblest
gift of man to imitate !

CHAPTER LI.

Description of Plate VII.

Number 1. FREDERICK OF PRUSSIA.

How much yet how little is there of the royal countenance in this copy! The covered forehead may be suspected from this nose, this sovereign feature. The forked,
descending wrinkles of the nose are expressive of killing
contempt. The great eyes, with a nose so bony, denote
a firmness and fire not easily to be withstood. Wit and
satirical fancy are apparent in the mouth, though defectively drawn. There is something minute seen in the
chin, which cannot well be in nature.

Number 2. CATHERINE, EMPRESS OF RUSSIA.

Except the smallness of the nostril, and the distance
of the eyebrow from the outline of the forehead, no one
can mistake the princely, the superior, the masculine
firmness of this, nevertheless feminine, but fortunate
and kind countenance.

Fredᵏ of Prussia.
Empˢˢ of Russia
Voltaire
Malherbe
Voisin
Lavater.

Number 3.   VOLTAIRE.

Precision is wanting to the outline of the eye, power to the eyebrows, the sting, the scourge of satire to the forehead. The under part of the profile, on the contrary, speaks a flow of wit, acute, exuberant, exalted, ironical, never deficient in reply.

Number 4.   F. DE MALHERBE.

Here is a high, comprehensive, powerful, firm, retentive French forehead, that appears to want the open, free, noble essence of the former; has something rude and productive; is more choleric; and its firmness appears to border on harshness.

Number 5.   J. DE VOISIN.

The delicate construction of the forehead, the aspect of the man of the world, the beauty of the nose, in particular, the somewhat rash, satirical mouth, the pleasure-loving chin, all show the Frenchman of a superior class. The excellent companion, the fanciful wit, the supple courtier, are every where apparent.

Number 6.   J. C. LAVATER.

A bad likeness of the author of these fragments, yet not to be absolutely mistaken. The whole aspect, especially the mouth, speaks inoffensive tranquillity, and benevolence bordering on weakness ;—more understanding and less sensibility in the nose than the author supposes himself to possess—some talent for observation in the eye and eyebrows.

CHAPTER LII.

Miscellaneous Quotations.

1.

" CAMPANELLA has not only made very accurate observations on human faces, but was very expert in mimicking such as were any way remarkable. Whenever he
thought proper to penetrate into the inclinations of those
he had to deal with, he composed his face, his gestures,
and his whole body, as nearly as he could, into the exact
similitude of the person he intended to examine, and
then carefully observed what turn of mind he seemed to
acquire by this change.   So that, says my author, he
was able to enter into the disposition and thoughts of
people, as effectually as if he had been changed into the
very man.   I have often observed that, on mimicking
the gestures and looks of angry, or placid, or frightened,
or daring men, I have involuntarily found my mind
turned to that passion whose appearance I endeavour to
imitate.   Nay, I am convinced it is hard to avoid it,
though one strove to separate the passion from its correspondent gestures.   Our minds and bodies are so closely
and intimately connected, that one is incapable of pain
or pleasure without the other.   Campanella, of whom
we have been speaking, could so abstract his attention
from any sufferings of his body, that he was able to
endure the rack itself without much pain ; and in lesser
pains every body must have observed that, when we can
employ our attention on any thing else, the pain has
been for a time suspended.   On the other hand, if by
any means the body is indisposed to perform such
gestures, or to be stimulated into such emotions as any
passion usually produces in it, that passion itself never

can arise, though its cause should be never so strongly in action, though it should be merely mental, and immediately affecting none of the senses. · As an opiate or spirituous liquor shall suspend the operation of grief, fear, or anger, in spite of all our efforts to the contrary; and this by inducing in the body a disposition contrary to that which it receives from these passions." This passage is extracted from Burke on the Sublime and Beautiful.

2.

"Who can explain wherein consists the difference of organization between an idiot and another man?"

The naturalist, whether Buffon or any other, who is become famous, and who can ask this question, will never be satisfied with any given answer, even though it were the most formal demonstration.

3.

"Diet and exercise would be of no use when recommended to the dying."

No human wisdom or power can rectify; but that which is impossible to man is not so to God.

4.

"The appearance without must be deformity and shame, when the worm gnaws within."

Let the hypocrite, devoured by conscience, assume whatever artful appearance he will, of severity, tranquillity, or vague solemnity, his distortion will ever be apparent to the physiognomist.

5.

"Take a tree from its native soil, its free air, and mountainous situation, and plant it in the confined circulation of a hothouse : there it may vegetate, but in a weak and sickly condition. Feed this foreign animal in a den; you will find it in vain. It starves in the midst of plenty, or grows fat and feeble."

This, I am sorry to say, is the mournful history of many a man.

6.

" A portrait is the ideal of an individual, not of men in general."

A perfect portrait is neither more nor less than the circular form of a man reduced to a flat surface, and which shall have the exact appearance of the person for whom it was painted seen in a camera obscura.

7.

I once asked a friend, " How does it happen that artful and subtle people always have one or both eyes rather closed ?"—"Because they are feeble," answered he. " Who ever saw strength and subtlety united ? The mistrust of others is meanness towards ourselves."

8.

This same friend, who to me is a man of ten thousand for whatever relates to mind, wrote two valuable letters on physiognomy to me, from which I am allowed to make the following extracts :—

" It appears to me to be an eternal law, that the first is the only true impression. Of this I offer no proof except by asserting such as my belief, and by appealing to the sensations of others. The stranger affects me by his appearance, and is to my sensitive being what the sun would be to a man born blind restored to sight.

9.

" Rousseau was right when he said of D., That man does not please me, though he has never done me any injury ; but I must break with him before it comes to that."

10.

" Physiognomy is as necessary to man as language." I may add, as natural.

CHAPTER LIII.

Miscellaneous Thoughts.

1.

EVERY thing is good. Every thing may, and must be misused. Physiognomical sensation is in itself as truly good, as godlike, as expressive of the exalted worth of human nature, as moral sensation ; perhaps they are both the same. The suppressing, the destroying a sensation so deserving of honour, where it begins to act, is sinning against ourselves, and in reality equal to resisting the good spirit. Indeed, good impulses and actions must have their limits, in order that they may not impede other good impulses and actions.

2.

Each man is a man of genius in his large or small sphere. He has a certain circle in which he can act with inconceivable force. The less his kingdom, the more concentrated is his power ; consequently the more irresistible is his form of government. Thus the bee is the greatest of mathematicians as far as its wants extend. Having discovered the genius of a man, how inconsiderable soever the circle of his activity may be, having caught him in the moment when his genius is in its highest exertion, the characteristic token of that genius will also be easily discovered.

3.

The approach of the Godhead cannot be nearer in the visible world, and in what we denominate nature, than in the countenance of a great and noble man. Christ could not but truly say, "He who seeth Me seeth Him that sent me." God cannot, without a miracle, be seen any where so fully as in the countenance of a good man. Thus the essence of any man is more present, more certain to me, by having obtained his shade.

4.

Great countenances awaken and stimulate each other, excite all that can be excited. Any nation, having once produced a Spenser, a Shakspeare, and a Milton, may be certain that a Steele, a Pope, and an Addison will follow. A great countenance has the credentials of its high original in itself. With calm reverence and simplicity nourish the mind with the presence of a great countenance; its emanations shall attract and exalt thee. A great countenance, in a state of rest, acts more powerfully than a common countenance impassioned; its effects, though unresembling, are general. The fortunate disciples, though they knew Him not, yet did their hearts burn within them while He talked with them by the way, and opened to them the scriptures. The buyers and sellers, whom he drove out of the Temple, durst not oppose Him.

It may from hence be conceived how certain persons, by their mere persons, have brought a seditious multitude back to their duty, although the latter had acquired the full power. That natural, unborrowed, indwelling power, which is consequently superior to any which can be assumed, is as evident to all eyes as the thunder of heaven is to all ears.

5.

Great physiognomical wisdom not only consists in discovering the general character of, and being highly affected by the present countenance, or this or that particular propensity, but in discriminating the individual character of each kind of mind, and its capacity, and being able to define the circle beyond which it cannot pass; to say what sensations, actions, and judgments are, or are not to be expected from the man under consideration, that we may not idly waste power, but dispense just sufficient to actuate and put him in motion.

No man is more liable to the error of thoughtless haste than I was. Four or five years of physiognomical observation were requisite to cure me of this too hasty waste of power. It is a part of benevolence to give, intrust, and participate; but physiognomy teaches when, how, and to whom to give. It therefore teaches true benevolence, to assist where assistance is wanted, and will be accepted. Oh! that I could call at the proper moment, and with proper effect, to the feeling and benevolent heart. Waste not, cast not thy seed upon the waters, or upon a rock. Speak only to the hearer; unbosom thyself but to those who can understand thee; philosophize with none but philosophers; spiritualize only with the spiritual. It requires greater power to bridle strength than to give it the rein. To withhold is often better than to give. What is not enjoyed will be cast back with acrimony or trodden to waste, and thus will become useless to all.

6.

To the good be good; resist not the irresistible countenance. Give the eye that asks, that comes recommended to thee by Providence, or by God himself, and which to reject is to reject God, who cannot ask thee more powerfully than when entreating in a cheerful, open, innocent countenance. Thou canst not more immediately glorify God than by wishing and acting well to a countenance replete with the spirit of God, nor more certainly and abhorrently offend and wound the majesty of God, than by despising, ridiculing, and turning from such a countenance. God cannot more effectually move man than by man. Whoever rejects the man of God rejects God. To discover the radiance of the Creator in the visage of man, is the pre-eminent quality of man; it is the summit of wisdom and benevolence to feel how much of this radiance is there, to discern this ray of Divinity through the clouds of the most debased countenances, and dig out this small gem of heaven from amid the ruins and rubbish by which it is encumbered.

7.

Shouldest thou, friend of man, esteem physiognomy as highly as I do, to whom it daily becomes of greater worth the more I discover its truth; if thou hast an eye to select the few noble, or that which is noble in the ignoble, that which is divine in all men, the immortal in what is mortal, then speak little, but observe much; dispute not, but exercise thy sensation, for thou wilt convince no one to whom this sensation is wanting.

When noble poverty presents to you a face in which humility, patience, faith, and love, shine conspicuously, how superior will thy joy be in his words who has told thee, " Inasmuch as thou hast done it unto one of the least of these my brethren, thou hast done it unto me ! "

With a sigh of hope you will exclaim, when youth and dissipation present themselves, This forehead was delineated by God for the search and the discovery of truth. In this eye rests unripened wisdom.

CHAPTER LIV.

Of the Union between the Knowledge of the Heart and Philanthropy.—Miscellaneous Physiognomical Thoughts from Holy Writ.

MAY the union between the knowledge of the heart and philanthropy be obtained by the same means? Does not a knowledge of the heart destroy or weaken philanthropy ? Does not our good opinion of any man diminish when he is perfectly known ? And if so, how may philanthropy be increased by this knowledge?

What is here alleged is truth; but it is partial truth. And how fruitful a source of error is partial truth! It is a certain truth that the majority of men are losers by being accurately known; but it is no less true that the majority of men gain as much on one side as they lose on the other by being thus accurately known. Who is so wise as never to act foolishly ? Where is the virtue wholly unpolluted by vice; with thoughts at all moments simple, direct, and pure? I dare undertake to maintain that all men, with some very rare exceptions, lose by being known. But it may also be proved by the most irrefragable arguments, that all men gain by being known; consequently a knowledge of the heart is not detrimental to the love of mankind, but promotes it.

Physiognomy discovers actual and possible perfections which, without its aid, must ever have remained hidden. The more man is studied, the more power and positive goodness will he be discovered to possess. As the experienced eye of the painter perceives a thousand small shades and colours which are unremarked by common spectators; so the physiognomist views a multitude of actual or possible perfections, which escape the general eye of the despiser, the slanderer, or even the more benevolent judge of mankind.

The good which I, as a physiognomist, have observed in people round me, has more than compensated that mass of evil which, though I appeared blind, I could not avoid seeing. The more I have studied man the more have I been convinced of the general influence of his faculties; the more have I remarked that the origin of all evil is good; that those very powers which made him evil—those abilities, forces, irritability, elasticity, were all in themselves actual, positive good. The absence of these, indeed, would have occasioned the absence of an infinity of evil, but so would they likewise of an infinity of good. The essence of good has given birth to much evil; but it contained in itself the possibility of a still infinite increase of good.

The least failing of an individual incites a general outcry, and his character is at once darkened, trampled on, and destroyed. The physiognomist views and praises the man whom the whole world condemns. What! does he praise vice?—Does he excuse the vicious?—No; he whispers, or loudly affirms, ".Treat this man after such a manner, and you will be astonished at what he is able, what he may be made willing to perform. He is not so wicked as he appears; his countenance is better than his actions. His actions, it is true, are legible in his countenance, but not more legible than his great powers, his sensibility, the pliability of that heart which has had an improper bent. Give but these powers which have rendered him vicious another direction and other objects, and he will perform miracles of virtue."

The physiognomist will pardon where the most benevolent philanthropist must condemn. For myself, since I have become a physiognomist, I have gained knowledge so much more accurate of so many excellent men, and have had such frequent occasions to rejoice my heart in the discoveries I made concerning such men, that this, as I may say, has reconciled me to the whole human race. What I here mention as having happened to myself, each physiognomist, being himself a man, must have undoubtedly felt.

Miscellaneous Physiognomical Thoughts from Holy Writ.

"Thou hast set our iniquities before thee, our secret sins in the light of thy countenance," Psalm xc. 8.— No man believes in the omniscience, or has so strong a conviction of the presence of God and his angels, or reads the hand of Heaven so visible in the human countenance, as the physiognomist.

" Which of you by taking thought can add one cubit unto his stature ?—And why take ye thought for rai- ment ?—Seek ye first the kingdom of God, and his righteousness ; and all these things shall be added unto you," Matt. vi. 27, 28, 33.—No man,.therefore, can alter his form. The improvement of the internal will also be the improvement of the external. Let men take care of the internal, and a sufficient care of the external will be the result.

"When ye fast, be not, as the hypocrites, of a sad countenance; for they disfigure their faces, that they may appear unto men to fast. Verily I say unto you, They have their reward. But thou, when thou fastest, anoint thine head and wash thy face; that thou appear not unto men to fast, but unto thy Father which is in secret: and thy Father which seeth in secret, shall reward thee openly," Matt. vi. 16—18. Virtue, like vice, may be concealed from men, but not from the Father in secret, nor from him in whom his spirit is, who fathoms not only the depths of humanity but of divinity. He is rewarded who means that the good he has should be seen in his countenance.

"Some seeds fell by the wayside, and the fowls came and devoured them up; some fell upon stony places, where they had not much earth: and forthwith they sprung up, because they had no deepness of earth; and when the sun was up, they were scorched; and because they had no root, they withered away; and some fell among thorns; and the thorns sprung up, and choked them; but others fell into good ground, and brought forth fruit, some a hundred-fold, some sixty-fold, some thirty-fold," Matt. xiii. 4—8. There are many men, many countenances, in whom nothing can be planted, each fowl devours the seed; or, they are hard like stone, with little earth, (or flesh,) have habits which stifle all that is good. There are others that have good bones, good flesh, with a happy proportion of each, and no stifling habits.

"For whosoever hath, to him shall be given, and he shall have more abundance; but whosoever hath

not, from him shall be taken away even that he hath,"
Matt. xiii. 12.—True again of the good and bad countenance. He who is faithful to the propensities of nature,
he hath, he enjoys, he will manifestly be ennobled. The
bad will lose even the good traits he hath received.

"Take heed that ye despise not one of these little
ones; for I say unto you, That in heaven their angels do
always behold the face of my Father which is in
heaven," Matt. xviii. 10.—Probably the angels see the
countenance of the father in the countenance of the
children.

"If any man have ears to hear, let him hear. Do ye
not perceive, that whatever thing from without entereth
into the man, it cannot defile him; because it entereth
not into his heart, but into the belly, and goeth out into
the draught, purging all meats? And he said, That
which cometh out of the man, that defileth the man,"
Mark vii. 17—20. This is physiognomically true. Not
external accidents, not spots which may be washed
away, not wounds which may be healed, not even scars
which remain, will defile the countenance in the eye of
the physiognomist, neither can paint beautify it to him.

"A little leaven leaveneth the whole lump," Gal. v. 9.
A little vice often deforms the whole countenance. One
single false trait makes the whole a caricature.

"Ye are our epistle, written in our hearts, known and
read of all men. Forasmuch as ye are manifestly
declared to be the epistle of Christ ministered by us,
written not with ink, but with the Spirit of the living
God," 2 Cor. iii. 2, 3. What need have the good of
letters of recommendation to the good? The open
countenance recommends itself to the open countenance.
No letters of recommendation can recommend the per-

fidious countenance, nor can any slanderer deprive the countenance, beaming with the divine spirit, of its letters of recommendation. A good countenance is the best letter of recommendation.

I shall conclude with the important passage from the eleventh of the Romans :

"God hath concluded them all in unbelief, that he might have mercy upon all. O the depth of the riches both of the wisdom and knowledge of God! How unsearchable are his judgments, and his ways past finding out! For who hath known the mind of the Lord? or who hath been his counsellor? or who hath first given to him, and it shall be recompensed unto him again? For of him, and through him, and to him, are all things. To whom be glory for ever. Amen."

CHAPTER LV.

Of the apparently false Decisions of Physiognomy—Of the general Objections made to Physiognomy—Particular Objections answered.

ONE of the strongest objections to the certainty of physiognomy is, that the best physiognomists often judge very erroneously.

It may be proper to make some remarks on this objection.

Be it granted that the physiognomist often errs ; that is to say his discernment errs, not the countenance. But to conclude there is no such science as physiognomy, because physiognomists err, is the same thing as to conclude there is no reason, because there is much false reasoning.

To suppose that, because the physiognomist has made some false decisions, he has no physiognomical discernment, is equal to supposing that a man who has committed some mistakes of memory, has no memory; or, at best, that his memory is very weak.—We must be less hasty. We must first inquire in what proportion his memory is faithful, how often it has failed, how often been accurate. The miser may perform ten acts of charity; must we therefore affirm he is charitable? Should we not rather inquire how much he might have given, and how often it has been his duty to give? The virtuous man may have ten times been guilty; but before he is condemned it ought to be asked, in how many hundred instances he has acted uprightly. He who games must oftener lose than he who refrains from gaming. He who slides or skates upon the ice is in danger of many a fall, and of being laughed at by the less adventurous spectator. Whoever frequently gives alms is liable occasionally to distribute his bounties to the unworthy. He indeed who never gives cannot commit the same mistake, and may truly vaunt of his prudence, since he never furnishes opportunities for deceit. In like manner, he who never judges can never judge safely. The physiognomist judges oftener than the man who ridicules physiognomy; consequently must oftener err than he who never risks a physiognomical decision.

Which of the favourable judgments of the benevolent physiognomists may not be decried as false? Is he not himself a mere man, however circumspect, upright, honourable, and exalted he may be; a man who has in himself the root of all evil, the germ of every vice; or, in other words, a man whose most worthy propensities,

qualities, and inclinations, may occasionally be over-
strained, wrested, and warped?

You behold a meek man who, after repeated and continued provocations to wrath, persists in silence; who
probably never is overtaken by anger when he himself
alone is injured. The physiognomist can read his heart,
fortified to bear and forbear, and immediately exclaims,
Behold the most amiable, the most unconquerable gentleness! You are silent—you laugh—you leave the place
and say, "Fie on such a physiognomist! How full of
wrath have I seen this man!"—When was it that you
saw him in wrath? Was it not when some one had mis-
treated his friend?—"Yes, and he behaved like a frantic
man in defence of his friend, which is proof sufficient
that the science of physiognomy is a dream, and the
physiognomist a dreamer." But who is in an error, the
physiognomist or his censurer?—The wise man may
sometimes utter folly. This the physiognomist knows,
but, regarding it not, reverses, and pronounces him a wise
man. You ridicule the decision, for you have heard this
wise man say a foolish thing.—Once more, who is in an
error? The physiognomist does not judge from a single
incident, and often not from several combining incidents.
Nor does he, as a physiognomist, judge only by actions.
He observes the propensities, the character, the essential
qualities and powers, which often are apparently con-
tradicted by individual actions.

Again; he who seems stupid or vicious may yet probably possess indications of a good understanding, and
propensities to every virtue. Should the beneficent eye
of the physiognomist, who is in search of good, perceive
these qualities and announce them; should he not pronounce a decided judgment against the man, he im-

mediately becomes a subject of laughter. Yet how often may dispositions to the most heroic virtue be there buried! How often may the fire of genius lie deeply smothered beneath the embers?—Wherefore do you so anxiously, so attentively, rake among these ashes? Because here is warmth, notwithstanding that at the first, second, third, fourth raking, dust only will fly in the eyes of the physiognomist and spectator. The latter retires laughing, relates the attempt, and makes others laugh also. The former may, perhaps, patiently wait and warm himself by the flame he has excited. Innumerable are the instances where the most excellent qualities are overgrown and stifled by the weeds of error. Futurity shall discover why, and the discovery shall not be in vain. The common unpractised eye beholds only a desolate wilderness. Education, circumstances, necessities, stifle every effort towards perfection. The physiognomist inspects, becomes attentive, and waits. He sees and observes a thousand contending contradictory qualities; he hears a multitude of voices exclaiming, What a man! But he hears too the voice of the Deity exclaim, What a man! He prays, while those revile who cannot comprehend, or, if they can, will not, that in the countenance, under the form they view, lie concealed beauty, power, wisdom, and a divine nature.

Still further, the physiognomist, or observer of man, who is a man, a Christian—that is to say, a wise and good man—will a thousand times act contrary to his own physiognomical sensation; I do not express myself accurately—he appears to act contrary to his internal judgment of the man. He speaks not all he thinks. This is an additional reason why the physiognomist so often appears to err; and why the true observer (observation and truth are in him) is so often mistaken and ridiculed. He reads the villain in the countenance of the beggar at his door, yet does not turn away, but speaks friendly to him, searches his heart, and discovers—O God, what does he discover!—An immeasurable abyss, a chaos of vice! But does he discover nothing more, nothing good? Be it granted he finds nothing good, yet he there contemplates clay which must not say to the potter, " Why hast thou made me thus ?" He sees, prays, turns away his face, and hides a tear which speaks with eloquence inexpressible, not to man, but to God alone. He stretches out his friendly hand, not only in pity to a hapless wife whom he has rendered unfortunate,—not only for the sake of his helpless, innocent children, but in compassion to himself, for the sake of God, who has made all things, even the wicked themselves, for his own glory. He gives, perhaps to kindle a spark which he yet perceives, and this is what is called in Scripture giving his heart. Whether the unworthy man misuses the gift, or misuses it not, the judgment of the donor will alike be arraigned. Whoever hears of the gift will say, How has this good man again suffered himself to be deceived !

Man is not to be the judge of man ; and who feels this truth more coercively than the physiognomist? The mightiest of men, the Ruler of man, came not to judge the world, but to save. Not that he did not see the vices of the vicious, nor that he concealed them from himself or others, when philanthropy required they should be remarked and detected; yet he judged not, punished not; he forgave—" Go thy way, and sin no more." Judas he received as one of his disciples, protected him, embraced him—him in whom he beheld his future betrayer.

Good men are most happy to discover good. Thine eye cannot be Christian if thou givest me not thine heart. Wisdom without goodness is folly; I will judge justly and act benevolently.

Once more. A profligate man, an abandoned woman, who have ten times been to blame when they affirmed they were not, on the eleventh are condemned when they are not to blame. They apply to the physiognomist. He inquires, and finds that this time they are innocent. Discretion loudly tells him he will be censured should he suffer it to be known that he believes them innocent; but his heart more loudly commands him to speak, to bear witness for the present innocence of such rejected persons. A word escapes him, and a multitude of reviling voices at once are heard —"Such a judgment ought not to have been made by a physiognomist!" Yet who has decided erroneously?

The above are a few hints and reasons to the discerning, to induce them to judge cautiously concerning the physiognomist, as they would wish him to judge concerning themselves or others.

Of the General Objections made to Physiognomy.

Innumerable are the objections which may be raised against the certainty of judgments drawn from the lines and features of the human countenance. Many of these appear to me to be easy, many difficult, and some impossible to be answered.

Before I select any of them, I will first state some general remarks, the accurate consideration and proof of which will remove many difficulties.

It appears to me that, in all researches, we ought first to inquire what can be said in defence of any proposition. One irrefragable proof of the actual existence and certainty of a thing, will overbalance ten thousand objections. One positive witness, who has all possible certainty that knowledge and reason can give, will preponderate against innumerable others who are only negative. All objections against a certain truth are in reality only negative evidence. "We never observed this : we never experienced that." Though ten thousand should make this assertion, what would it prove against one man of understanding and sound reason who should answer, "But I have observed, and you also may observe if you please." No well-founded objection can be made against the existence of a thing visible to sense. Argument cannot disprove facts. No two opposing positive facts can be adduced; all objections to a fact, therefore, must be negative.

Let this be applied to physiognomy. Positive proofs of the true and acknowledged signification of the face and its features, against the clearness and certainty of which nothing can be alleged, render innumerable objections, although they cannot probably be answered, perfectly insignificant. Let us therefore endeavour to inform ourselves of those positive arguments which physiognomy affords. Let us first make ourselves steadfast in what is certainly true, and we shall soon be enabled to answer many objections, or to reject them as unworthy any answer.

It appears to me, that in the same proportion as a man remarks and adheres to the positive, will be the strength and perseverance of his mind. He whose talents do not surpass mediocrity, is accustomed to overlook the positive, and to maintain the negative with invincible obstinacy.

Thou shouldest first consider what thou art, what is thy knowledge, and what are thy qualities and powers, before thou inquirest what thou art not, knowest not, and what the qualities and powers are that thou hast not. This is a rule which every man who wishes to be wise, virtuous, and happy, ought not only to prescribe to himself, but, if I may use so bold a figure, to incorporate with, and make a part of his very soul. The truly wise always first directs his inquiries concerning what is; the man of weak intellect, the pedant, first searches for that which is wanting. The true philosopher looks first for the positive proofs of the proposition. I say, first—I am very desirous that my meaning should not be misunderstood, and therefore repeat, *first*, The superficial mind first examines the negative objections. This has been the method pursued by infidels, the opponents of Christianity. Were it granted that Christianity were false, still this method would neither be logical, true, nor conclusive. Therefore such modes of reasoning must be set aside as neither logical nor conclusive, before we can proceed to answer objections.

To return once more to physiognomy, the question will be reduced to this—" Whether there are any proofs sufficiently positive and decisive, in favour of physiognomy, to induce us to disregard the most plausible objections ?"—Of this I am as much convinced as I am of my own existence ; and every unprejudiced reader will be the same who shall read this work through, if he only possesses so much discernment and knowledge as not to deny that eyes are given us to see; although there are innumerable eyes in the world that look and do not see.

It may happen that learned men of a certain description will endeavour to perplex me by argument. They, for example, may cite the female butterfly of Réaumur, and the large-winged ant, in order to prove how much we may be mistaken with respect to final causes in the products of nature. They may assert, " Wings undoubtedly appear to be given for the purpose of flight, yet these insects never fly ; therefore wings are not given for that purpose. And, by a parity of reasoning, since there are wise men who probably do not see, eyes are not given for the purposes of sight."—To such objections I shall make no reply, for never in my whole life have I been able to answer·a sophism. I appeal only to common sense. I view a certain number of men, who all have the gift of sight when they open their eyes and there is light, and who do not see when their eyes are shut. As this certain number are not select, but taken promiscuously among millions of existing men, it is the highest possible degree of probability that all men whose formation is similar, that have lived, do live, or shall live, being alike provided with those organs we call eyes, must see. This, at least, has been the mode of arguing and concluding among all nations, and in all ages. In the same degree as this mode of reasoning is convincing, when applied to other subjects, so it is when applied to physiognomy, and is equally applicable; and, if untrue in physiognomy, it is equally untrue in every other instance.

I am therefore of opinion that the defender of physiognomy may rest the truth of the science on this proposition, " That it is universally confessed that among ten, twenty, or thirty men, indiscriminately selected, there as certainly exists a physiognomical expression, or demonstrable correspondence of internal power and sensation, with external form and figure, as that among the like number of men, in the like manner selected, they have eyes and can see." Having proved this, he has as sufficiently proved the universality and truth of physiognomy as the universality of sight by the aid of eyes, having shown that ten, twenty, or thirty men, by the aid of eyes, are all capable of seeing. From a part I draw a conclusion to the whole; whether those I have seen or those I have not.

But it will be answered, though this may be proved of certain features, does it therefore follow that it may be proved of all?—I am persuaded it may; if I am wrong, show me my error.

Having remarked that men who have eyes and ears see and hear, and being convinced that eyes were given him for the purpose of sight, and ears for that of hearing; being unable longer to doubt that eyes and ears have their destined office—I think I draw no improper conclusion, when I suppose that every other sense and member of this same human body, which so wonderfully form a whole, has each a particular purpose; although it should happen that I am unable to discover what the particular purposes of so many senses, members, and integuments may be. Thus do I reason, also, concerning the signification of the countenance of man, the formation of his body, and the disposition of his members.

If it can be proved that any two or three features have a certain determinate signification, as determinate as that the eye is the expression of the countenance, is it not accurate to conclude, according to the mode of reasoning above cited, universally acknowledged to be just, that those features are also significant, with the signification of which I am unacquainted I think myself able to prove, to every person of the commonest understanding, that all men without exception, at least under certain circumstances and in some particular feature, may indeed have more than one feature of a certain determinate signification, as surely as I can render it comprehensible to the simplest person, that certain determinate members of the human body are to answer certain determinate purposes.

Twenty or thirty men taken promiscuously, when they laugh or weep, will, in the expression of their joy or grief, possess something in common with or similar to each other. Certain features will bear a greater resemblance to each other among them than they otherwise do, when not in the like sympathetic state of mind.

To me it appears evident, that since excessive joy and grief are universally acknowledged to have their peculiar expressions, and that the expression of each is as different as the different passions of joy and grief, it must therefore be allowed that the state of rest, the medium between joy and grief, shall likewise have its peculiar expression ; or, in other words, that the muscles which surround the eyes and lips will indubitably be found to be in a different state.

If this be granted concerning the state of the mind in joy, grief, or tranquillity, why should not the same be true concerning pride, humility, patience, magnanimity, and other affections ?

According to certain laws, the stone flies upward when thrown with sufficient force ; by other laws equally certain, it afterwards falls to the earth ; and will it not remain unmoved according to laws equally fixed, if suffered to be at rest ? Joy, according to certain laws, is expressed in one manner, grief in another, and tranquillity in a third. Wherefore then shall not anger, gentleness, pride, humility, and other passions, be subject to certain laws; that is, to certain fixed laws?

All things in nature are or are not subjected to certain laws. There is a cause for all things, or there is not. All things are cause and effect, or are not. Ought we not hence to derive one of the first axioms of philosophy? And, if this be granted, how immediately is physiognomy relieved from all objections, even from those which we know not how to answer; that is, as soon as it shall be granted there are certain characteristic features in all men, as characteristic as the eyes are to the countenace!

But, it will be said, how different are the expressions of joy and grief, of the thoughtful and the thoughtless! And how may these expressions be reduced to 'rule?

How different from each other are the eyes of men and of all creatures—the eye of an eagle from the eye of a mole, an elephant, and a fly! and yet we believe of all who have no evident signs of infirmity or death, that they see.

The feet and ears are as various as are the eyes; yet we universally conclude of them all, they were given us for the purposes of hearing and walking.

These varieties by no means prevent our believing that the eyes, ears, and feet, are the expressions, the organs of seeing, hearing, and walking; and why should we not draw the same conclusions concerning all features and lineaments of the human body? The expressions of similiar dispositions of mind cannot have greater variety than have the eyes, ears, and feet of all beings that see, hear, and walk; yet may we as easily observe and determine what they have in common, as we can observe and determine what the eyes, ears, and feet, which are so various among all beings that see, hear, and walk, have also in common. This well considered, how many objections will be answered, or become insignificant!

Various Objections to Physiognomy answered.

Objection 1.

" It is said we find persons who from youth to old age, without sickness, without debauchery, have continually a pale, death-like aspect; who, nevertheless, enjoy an uninterrupted and confirmed state of health."

Answer.

These are uncommon cases. A thousand men will show their state of health by the complexion and roundness of the countenance, to one in whom these appearances will differ from the truth. I suspect that these uncommon cases are the effect of impressions made on the mother during her state of pregnancy. Such cases may be considered as exceptions, the accidental causes of which may, perhaps, not be difficult to discover.

To me it seems we have as little just cause hence to draw conclusions against the science of physiognomy, as we have against the proportion of the human body, because there are dwarfs, giants, and monstrous births.

Objection 2.

A friend writes me word, " He is acquainted with a man of prodigious strength, who, the hands excepted, has every appearance of weakness, and would be supposed weak by all to whom he should be unknown."

Answer.

I could wish to see this man. I much doubt whether his strength be only expressed in his hands, or, if it were, still it is expressed in the hands; and, were no exterior signs of strength to be found, still he must be considered as an exception, an example unexampled. But, as I have said, I much doubt the fact. I have never yet seen a strong man whose strength was not discoverable in various parts.

Objection 3.

" We perceive the signs of bravery and heroism in the countenances of men, who are, notwithstanding, the first to run away."

Answer.

The less the man is, the greater he wishes to appear.

But what were these signs of heroism? Did they resemble those found in the Farnesian Hercules ?—Of this I doubt: let them be drawn, let them be produced; the physiognomist will probably say, at the second, if not at the first glance, *quanta species!* Sickness, accident, melancholy, likewise deprive the bravest men of courage. This contradiction, however, ought to be apparent to the physiognomist.

Objection 4.

" We find persons whose exterior appearance denotes extreme pride, and who in their actions never betray the least symptom of pride."

Answer.

A man may be proud and affect humility.

Education and habit may give an appearance of pride, although the heart be humble; but this humility of heart will shine through an appearance of pride, as sunbeams through transparent clouds. It is true that this apparently proud man would have more humility had he less the appearance of pride.

Objection 5.

"We see mechanics who, with incredible ingenuity, produce the most curious works of art, and bring them to the greatest perfection; yet who, in their hands and bodies, resemble the rudest peasants and woodcutters; while the hands of fine ladies are totally incapable of such minute and curious performances."

Answer.

I should desire these rude and delicate frames to be brought together and compared. Most naturalists describe the elephant as gross and stupid in appearance; and according to this apparent stupidity, or rather according to that stupidity which they ascribe to him, wonder at his address. Let the elephant and the tender lamb be placed side by side, and the superiority of address will be visible from the formation and flexibility of the body, without further trial.

Ingenuity and address do not so much depend upon the mass as upon the nature, mobility, internal sensation, nerves, construction, and suppleness of the body and its parts.

Delicacy is not power; power is not minuteness.

Apelles would have drawn better with charcoal than
many miniature painters with the finest pencil. The
tools of a mechanic may be rude, and his mind the very
reverse. Genius will work better with a clumsy hand
than stupidity with a hand the most pliable. I will,
indeed, allow your objection to be well-founded, if
nothing of the character of an artist is discoverable in
his countenance; but, before you come to a decision, it
is necessary you should be acquainted with the various
marks that denote mechanical genius in the face. Have
you considered the lustre, the acuteness, the penetration
of his eyes; his rapid, his decisive, his firm aspect; the
projecting bones of his brow, his arched forehead, the
suppleness, the delicacy, or the massiness of his limbs ?
Have you well considered these particulars ? " I could
not see it in him," is easily said. More consideration is
requisite to discover the character of the man.

Objection 6.

" There are persons of peculiar penetration who have
very unmeaning countenances."

Answer.

The assertion requires proof.

For my own part, after many hundred mistakes, I
have continually found the fault was in my want of
proper observation. At first, for example, I looked for
the tokens of any particular quality too much in one
place; I sought and found it not, although I knew the
person possessed extraordinary powers. I have been
long before I could discover the seat of character. I was
deceived, sometimes by seeking too partially, at others
too generally. To this I was particularly liable in ex-

amining those who had only distinguished themselves in some particular pursuit; and, in other respects, appeared to be persons of very common abilities, men whose powers were all concentrated to a point, to the examination of one subject; or men whose powers were very indeterminate: I express myself improperly, powers which had never been excited, brought into action. Many years ago I was acquainted with a great mathematician, the astonishment of Europe; who at the first sight, and even long after, appeared to have a very common countenance. I drew a good likeness of him, which obliged me to pay a more minute attention, and found a particular trait which was very marking and decisive. A similar trait to this I, many years afterward, discovered in another person who, though widely different, was also a man of great talents; and who, this trait excepted, had an unmeaning countenance, which seemed to prove the science of physiognomy all erroneous. Never since this time have I discovered that particular trait in any man who did not possess some peculiar merit, however simple his appearance might be.

This proves how true and false, at once, the objection may be which states, "Such a person appears to be a weak man, yet has great powers of mind."

I have been written to concerning D'Alembert, whose countenance, contrary to all physiognomical science, was one of the most common. To this I can make no answer unless I had seen D'Alembert. This much is certain, that his profile by Cochin, which yet must be very inferior to the original, not to mention other less obvious traits, has a forehead, and in part a nose, which were never seen in the countenance of any person of moderate, not to say mean, abilities.

Objection 7.

"We find very silly people with very expressive countenances."

Who does not daily make this remark? My only answer, which I have repeatedly given, and which I think perfectly satisfactory, is, that the endowments of nature may be excellent; and yet by want of use, or abuse, may be destroyed. Power is there, but it is power misapplied; the fire wasted in the pursuit of pleasure can no longer be applied to the discovery and display of truth—it is fire without light, fire that ineffectually burns.

I have the happiness to be acquainted with some of the greatest men in Germany and Switzerland; and I can upon my honour assert, that of all the men of genius with whom I am acquainted, there is not one who does not express the degree of invention and powers of mind he possesses in the features of his countenance, and particularly in the form of his head.

I shall only select the following names from an innumerable multitude. Charles XII., Louis XIV., Turenne, Sully, Polignac, Montesquieu, Voltaire, Diderot, Newton, Clarke, Maupertuis, Pope, Locke, Swift, Lessing, Bodmer, Sultzer, Haller. I believe the character of greatness in these heads is visible in every well-drawn outline. I could produce numerous specimens, among which an experienced eye would scarcely ever be mistaken.

M'CORQUODALE AND CO., PRINTERS, LONDON—WORKS, NEWTON.